International Cross-Industry Safety Conference - International Symposium on Aircraft Technology, MRO and Operations (ICSC-ISATECH 2019)

MATEC Web of Conferences Volume 314 (2020)

Amsterdam, the Netherlands
9 – 11 October 2019

Editors:

Konstantinos Stamoulis
Maria Papanikou
Asteris Apostolidis
Anastasios Plioutsias
Hikmet Karakoc

ISBN: 978-1-7138-1290-6

Printed from e-media with permission by:

Curran Associates, Inc.
57 Morehouse Lane
Red Hook, NY 12571

Some format issues inherent in the e-media version may also appear in this print version.

For additional information, please contact EDP Sciences – Web of Conferences at the address below.

EDP Sciences – Web of Conferences
17, Avenue du Hoggar
Parc d'Activité de Courtabœuf
BP 112
F-91944 Les Ulis Cedex A
France

Phone: +33 (0) 1 69 18 75 75
Fax: +33 (0) 1 69 28 84 91

contact-edps@webofconferences.org

Additional copies of this publication are available from:

Curran Associates, Inc.
57 Morehouse Lane
Red Hook, NY 12571 USA
Phone: 845-758-0400
Fax: 845-758-2633
Email: curran@proceedings.com
Web: www.proceedings.com

TABLE OF CONTENTS

Amsterdam University of Applied Sciences

SARES

International Cross-Industry Safety Conference (ICSC) and International Symposium on Aircraft Technology, MRO and Operations (ISATECH) 2019

Editorial of Proceedings

Konstantinos Stamoulis[1], Maria Papanikou[1], Asteris Apostolidis[2] Anastasios Plioutsias[1,3] and Hikmet Karakoc[4]

[1]Aviation Academy, Amsterdam University of Applied Sciences, the Netherlands

[2]KLM Royal Dutch Airlines, the Netherlands

[3]School of Mechanical, Aerospace and Automotive Engineering, Faculty of Engineering, Environment and Computing, Coventry University, UK

[4]Eskisehir Technical University, TR

The Amsterdam University of Applied Sciences (AUAS) and the Sustainable Aviation Research Society (SARES) organised in Amsterdam (9-11 October 2019) the 4th edition of the International Cross-industry Safety Conference (ICSC) and the 2nd edition of the International Symposium on Aircraft Technology, MRO and Operations (ISATECH) dedicated to the interface between research and practice.

The AUAS Aviation Academy performs research into real-life cases and problems in the aviation sector, focusing on safety and human factors, MRO process improvement, composites & advanced maintenance technologies and aviation capacity. The overarching concept of the AA research is the Sustainable Operational Readiness; the development of sustainable processes & operations in Aviation. Our goal is to improve and innovate professional practice. We perform all of our research projects in close cooperation with industry, governmental agencies and scientific institutions and universities. This ensures a solid connection with state-of-the-art scientific knowledge, as well as a focus on the most current and urgent problems. The aim of this gathering is dual; to act as a forum, where innovative ideas and technologies are presented and to form the hub for the dissemination of multi-disciplinary knowledge and expertise between industry and academia.

ICSC addressed a variety of topics that center safety and human factors across industries, such as aviation and rail. Specifically, safety and risk management, systems thinking and safety culture, neuroergonomics, peer pilot support, safety investigations, and safety performance were discussed. The interdisciplinary thematic forms a commotional context in safety and human factors, calling industries and academia to an open dialogue. ISATECH served as a platform for the dissemination of multidisciplinary knowledge, innovative ideas and experience between industry and academia. ISATECH addressed a variety of topics such as advanced maintenance technologies, air transport operations, aerospace advanced materials, airport capacity, advances in composites, maintenance planning & scheduling, predictive maintenance, electric flight, propulsion technologies for aerospace vehicles, structural health monitoring systems, and sustainable aviation.

Maarten Koopmans (Vice President, Component Services, KLM) opened the conferences in a joined plenary session. Robert Jan de Boer (Director of Northumbria University in Amsterdam, NL), Stathis Malakis (Air Traffic Controller, Hellenic Civil Aviation Authority, GR), Colin Dennis

(Retired Technical Director of RSSB, UK), Tamara Pejovic (Safety Performance Expert, Eurocontrol, BE), David Lindley (Head of Aviation Safety and Quality Assurance at Hybrid Air Vehicles, UK), Umit Kus (Training Manager at myTechnik, TR), Arnold Gad-Briggs (Director of EGB Engineering, UK), Birol Kilkis (ASHRAE Fellow, Polar Technology Hacettepe University & European Technology and Innovation Platform, TR), Ravi Rajamani (Visiting Professor Cranfield University, UK & SAE Fellow, USA), and Kateryna Synylo (Assoc. Professor, Faculty of Environmental Safety, UA) addressed the conferences as keynote speakers. Besides the keynote speakers, 46 delegates from various industry, academia and (inter)governmental organisations delivered their presentations and discussed with the attendants a wide range of practical applications and research results.

We would like to thank the organising committee members, Rianne Boute, Renée Groen and Shen van Dijk under the leadership of Sanne van Dorp and Viktoria Balla-Kamper. We also express our deep appreciation to the programme committee members for their contribution in the review process (in alphabetical order of the last name):

International Cross-industry Safety Conference (ICSC)

- Ioannis Anyfantis, European University Cyprus- CERIDES, CY
- Robert-Jan de Boer, Northumbria University, UK
- Mikela Chatzimichailidou, WSP, UK
- Steve Denniss, WSP, UK
- Utku Kale, Budapest University of Technology and Economics, HU
- Nektarios Karanikas, Queensland University of Technology, AU
- Michalis Karyotakis, Hellenic Air Force, GR
- Dimitris Nathanael, National Technical University of Athens, GR
- Alfred Roelen, Netherlands Aerospace Centre, NL
- Antony Smoker, Lund University, SE
- Panagiotis Sotiriadis, National Center for Public Administration and Local Government, GR
- Sriramula Srinivas, University of Aberdeen, UK
- John Stoop, Kindynos, NL
- Tim Wincher, WSP, UK

International Symposium on Aircraft Technology, MRO and Operations (ISATECH)

- Maaik Boorst, Amsterdam University of Applied Sciences, NL
- Amare Desalegn Fentaye, Malardalen University, SE
- Alper Dalkiran, Suleyman Demirel University, TR
- Dimitra-Eirini Diamantidou, Malardalen University, SE
- Roberto Felix Patron, Amsterdam University of Applied Sciences, NL
- Aziz Kaba, Eskisehir Technical University, TR
- Hikmet Karakoc, Eskisehir Technical University, TR
- Alejandro Murrieta-Mendoza, Amsterdam University of Applied Sciences, NL
- Rogier Nijssen, Inholland University, NL
- Jozsef Rohacs, Budapest University of Technology and Economics, HU
- Onder Turan, Eskisehir Technical University, TR
- Melih Yildiz, Igdir University, TR
- Oleksandr Zaporozhets, National Aviation University, UA
- Catya Zuniga Alcaraz, Amsterdam University of Applied Sciences, NL

Statement of Peer review

In submitting conference proceedings to MATEC Web of Conferences, I certify to the Publisher that I adhere to the **Policy on Publishing Integrity** of the journal in order to safeguard good scientific practice in publishing.

1. All articles have been subjected to peer review administered by the proceedings editors.
2. Reviews have been conducted by expert referees, who have been requested to provide unbiased and constructive comments aimed, whenever possible, at improving the work.
3. Proceedings editors have taken all reasonable steps to ensure the quality of the materials they publish and their decision to accept or reject a paper for publication has been based only on the merits of the work and the relevance to the journal.

Title, date and place of the conference

International Cross-Industry Safety Conference (ICSC) and
International Symposium on Aircraft Technology, MRO and Operations (ISATECH) 2019

9-11 October

Amsterdam, NL

Proceedings editor(s): Konstantinos Stamoulis, Maria Papanikou, Asteris Apostolidis, Anastatios Plioutsias, and Hikmet Karakoc

Date and editor's signature

20.05.2020

A novel real-time safety level calculation approach based on STPA

Apostolos Zeleskidis[1,*], *Ioannis M.* Dokas[2], and *Basil* Papadopoulos[3]
[1]Undergraduate Student (Democritus University of Thrace, Civil Engineering, Greece)
[2]Assistant Professor (Democritus University of Thrace, Civil Engineering, Greece)
[3]Professor (Democritus University of Thrace, Civil Engineering, Greece)

Abstract. This paper proposes a novel approach to dynamic safety level calculation for safety-critical systems based on the STAMP accident model and the implementation of a mathematical model. The proposed approach utilises (1) an STPA hazard analysis applied to the system in question, (2) system operational data from domain experts regarding process duration and reaction times, and (3) real-time system data. The STPA analysis is transformed into acyclic diagrams that graphically indicate every possible sequence of safety constraint violations that could lead to a loss (path). Based on this diagram the safety level (SL) of a system is defined as SL=$\vec{p}_w$ where $\vec{p}_w$ is the most detrimental to safety path which is active for any possible time value or context in the system's operation. This approach is also demonstrated using as a case study the "classical" Train Door STPA analysis example. This paper aims to introduce a new perspective on the problem of measuring and managing the actual safety level of highly complex socio-technical systems in real time and discusses related limitations and future research opportunities of this approach.

1 Introduction

Several safety indicators are utilised to monitor safety drifts and to assess whether a system maintains its safety within acceptable levels. As a result, numerous categories of safety indicators have been proposed in the literature such as event indicators, barrier indicators, activity indicators, and programming indicators [1,2]. However, the need to "get smarter at predicting the next accident" [3] and the challenge of measuring what the actual safety level of a system is at a certain moment in time at a given context [4] still remains, regardless of the introduction of new accident models and the view of safety as an emergent system property. To address this problem, Leveson [5] proposed that useful leading indicators can be identified based on the assumptions [6] underlying our safety engineering practices rather than on the likelihood of loss events. Thus, there is a need to monitor whether the generated safety assumptions based on which a system was designed, hold during the phase of operations are vulnerable or changed. Chatzimicailidou et al. [7] suggested to measure the gap between system design and operation as a measure for safety and introduced the

* Corresponding author: aposzele@civil.duth.gr

RiskSOAP indicator. With RiskSOAP, one can compare the safety constraints of existing systems compared to their ideal set of safety constraints that derive from their STPA [8] and EWaSAP [9] analysis, and calculate a situational awareness indicator for the system under study.

The problem is that there are no methodologies paired with a mathematical model that can determine dynamically what the safety level is at any certain moment in time for any given context and how much time is left until an accident takes place. To fill this gap, we propose a novel approach including a mathematical model for systems' safety level determination and its dynamic calculation based on the STAMP accident model. The proposed model utilizes the outcomes of STPA hazard analyses which are transformed into diagrams. In these diagrams, each node depicts an STPA based safety constraint, and each path of the acyclic diagram depicts a possible scenario of safety constraints violations that can lead to accidents.

2 STPA

STPA is a hazard analysis technique based on the STAMP accident model. According to STAMP, safety is an emergent property of systems, and accidents can occur not just because of component failures but also due to unsafe interactions among system components that did not fail [8]. That means that the feedback control loops of the system should be designed, developed and operated in a manner that their controllers will not enforce unsafe control actions in any possible operational context or environment due to lack of awareness of the system state (i.e. mental model of the system). This could be achieved by applying the STPA hazard analysis during the life cycle of the system, ideally as early as possible.

The STPA hazard analysis consists of four steps. In the first step, the purpose of the analysis and the boundaries of the system are defined, together with the losses or accidents (A), the system level hazards (H), and the system level safety constraints. During the second step of STPA, a functional diagram known as the safety control structure diagram is developed, where the controllers and the controlled processes of the system are shown, together with the control actions and the feedback each controller is receiving from the process it controls.

In the third step, the control actions of each controller are analyzed to examine under which contexts they could lead to the identified losses. The so-called Unsafe Control Actions (UCA) are then translated into safety constraints or safety specifications. In addition, all Control Actions (CA) given by the controller are analysed to determine how they could be unable to affect the controlled process as intented while enforced by the controller within the appropriate context. The fourth step aims at identifying the reasons why unsafe control actions might be enforced in the system. As a result, loss scenarios (S) are created to explain how incorrect feedback, inadequate requirements, design errors, component failures and other factors that could cause unsafe or ineffective control actions (CA) which are grouped and analysed fundementaly as UCAs [8].

This analysis incorporates safety barriers and other similar mechanisms as parts of the system and finds the weaknesses in the complete system together with its defenses.

3 Methodology

The complete approach we suggest is depicted in Figure 1. The proposed model is comprised of two main processes. The results of STPA and EWaSAP analyses as well as time ranges taken from domain experts are used for the first step of the methodology which is the Model Development process of the system under study which leades to the creation of a mathematical model dedicated to the system.

Then, from real time system data and the operational mode of the system the Safety Level Determination process is initiated. Operational modes are a tool used in complicated systems to distinguish between different modes in a systems operation. For example, an airplane during a typical flight goes through 3 different operational modes: take off, landing and cruising. The goal of the Safety Level Determination process is to dynamically calculate the Safety Level of the system in real time

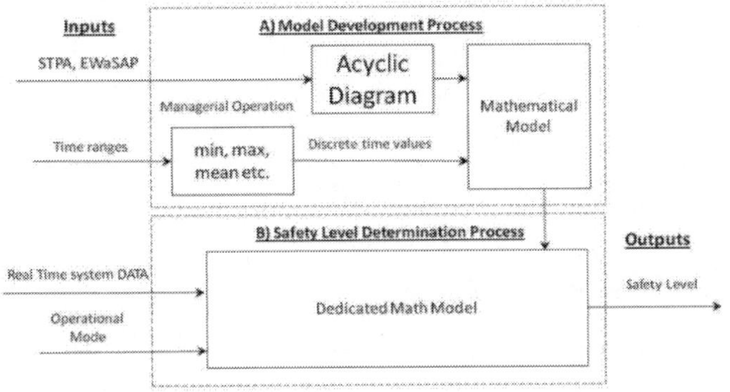

Figure. 1. The proposed approach.

3.1 Model Developments

In more detail, the main goal of the Model Development process is to create and populate an acyclic diagram that represents the system in terms of safety and calibrate the mathematical model to the exact system in study. This is achieved by utilising the following inputs:

1. Safety specifications and constraints obtained from the STPA analysis are turned into nodes of the diagram and are categorised according to their impact on safety. The Categories (levels) are ranked from most to least consequential according to STPA: Level 4 = Accidents, Level 3 = Hazards, Level 2 = Unsafe Control Actions (UCAs) and ineffective Control Actions (CA) and Level 1 = Scenarios. Also, safety constraints are causally linked as shown in the diagram by the connections between the nodes which represent the specifications. These connections are always between nodes of two ascending levels (from level 1 to level 2 or from level 2 to level 3 etc.). See Figure 2 for nodes connections and levels of safety constraints in acyclic diagrams.
2. Time ranges given by domain experts or observed by studying the system regarding the connections of the casually linked safety constraints. These should be distilled in the form of sets (min(t), max(t)) for every connection between two nodes, such as the time it would take from the moment that an Unsafe Control Action is given to the moment that a Hazardous state occurs because of the specific UCA in the system. Each set represents the maximum and minimum amount of time it would take for an active node to activate the node connected with it (Figure 3).
3. The operation used (minimum, maximum, mean average etc.) to discern a single time value $t_{x \to y}$ from the time ranges which represent the flexibility in how the entire operation can be managed. A path is defined as a possible scenario of safety constraints violations that can lead to accidents. It is one of the possible "roads" leading from the lowest level (loss scenarios) to the highest (losses) using the connections between nodes. For instance, the example of Figure 2 has a total of 17 paths and one possible path is S1→ UCA1→ H-2→ A (shown with bold arrows). Each path is comprised of 4 nodes,

one in each level of the tree-like structure. Since a path is comprised of 4 nodes it also has 3 corresponding time values ($t1\rightarrow2$, $t2\rightarrow3$, $t3\rightarrow4$). Paths have two main characteristics: i) Path completeness y which expresses which Level of the acyclic diagram is the highest active node of the path in time t. Path completeness takes values in the discrete set {1,2,3,4}, ii) Time remaining until accident $t+(y)$ which is calculated according to the path completeness and the singular time values assigned to each path and expresses the time which is left for a path to reach the node that represents the occurrence of an accident. An example of path completeness and time remaining until accident is given in Figure 3.

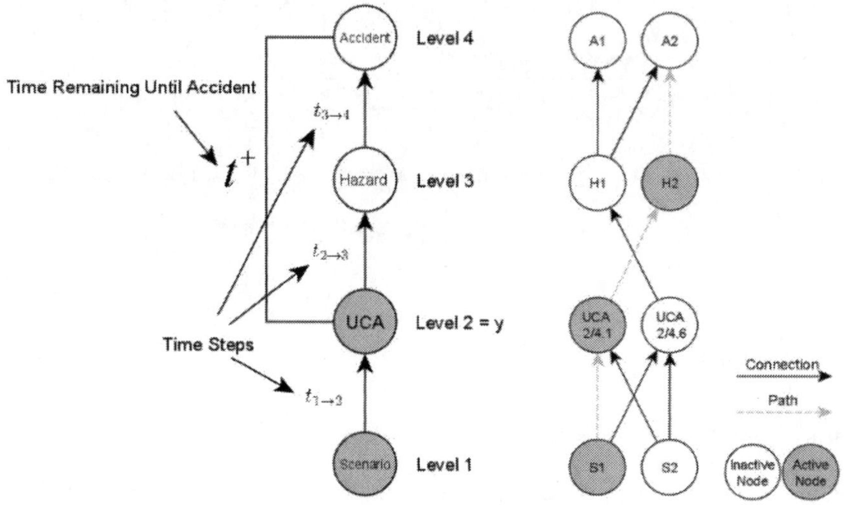

Figure. 2. (Right) Example of an acyclic diagram and representations of active and inactive nodes, paths, and node connections.

Figure. 3. (Left) Example path, completeness levels and the corresponding time values.

3.2 Safety Level Determination

The goal of the Safety Level Determination process is the actual real-time safety monitoring of the system during its operation. The data needed are:

1. The Operational Mode of the system. For instance, referring to a plane system during its take-off mode, out of the total set of the acyclic diagram paths that can be produced from the plane system, the paths which will be computed during the Operational System process of the methodology will be only those belonging to the take-off mode of plane operations.
2. System specific data according to the STPA analysis. This data must be able to show if any node or safety constraint of the diagram has activated or not for any time *t*. Activated means the safety constraint was violated in the specific time value monitored.

Using the mathematical model and the active nodes, the most detrimental to safety system path is calculated.

3.3 The mathematical model

The mathematical model defines the safety level of the system in a specific time t as $SL = \vec{p}_w(t) = \max(\vec{\wp})$, where $\vec{\wp}$ is the ordered set of all paths of the system derived by STPA, $\vec{p}_w(t)$ indicates the path or set of paths that for time t are considered the "worst" among all the paths of the diagram concerning their completeness level (i.e. how "close" to the loss node is the last active node of this path), and the time left until the accident (i.e. the time until the loss node is activated meaning an accident has taken place in the system) graphically represented in Figure 3.
Mathematically paths are defined as follows:
For path $\vec{p} = (y, t_{\vec{p}}^{+}(y)) \in Y \times \mathbb{R}^{+}$where:
y the path completeness for p in time t, y $\in \{0,1,2,3,4\}$
$t_{\vec{p}}^{+}(y)$:$\{0,1,2,3,4\} \rightarrow [0,\infty)$ is the time remaining until the accident of path p is activated and is calculated according to the following function:

$$t_{\vec{p}}^{+}(y) = \begin{cases} t_{1\to2}^{\vec{p}} + t_{2\to3}^{\vec{p}} + t_{3\to4}^{\vec{p}} & y \in \{0,1\} \\ t_{2\to3}^{\vec{p}} + t_{3\to4}^{\vec{p}} & y = 2 \\ t_{3\to4}^{\vec{p}} & y = 3 \\ 0 & y = 4 \end{cases} \quad (1)$$

We consider the set of all the paths $\vec{\wp} = \{\vec{p}_1, \vec{p}_2, \dots, \vec{p}_n\}$ where n is the number of paths of the complete system.
In the set $\vec{\wp}$ we define the following ordering relation, according to which $\max(\vec{\wp})$ is determined:

$$\forall \vec{p}_1\left(y_{\vec{p}_1}, t_{\vec{p}_1}^{+}(y_{\vec{p}_1})\right), \vec{p}_2\left(y_{\vec{p}_2}, t_{\vec{p}_2}^{+}(y_{\vec{p}_2})\right) \in \vec{\wp}$$
$$\vec{p}_1 \geq \vec{p}_2 \Leftrightarrow t_{\vec{p}_1}^{+} < t_{\vec{p}_2}^{+} \vee \left(t_{\vec{p}_1}^{+} = t_{\vec{p}_2}^{+} \wedge y_{\vec{p}_1} \geq y_{\vec{p}_2}\right) \quad (2)$$

4 Case study

The fictitious system of the train door STPA analysis example [10] will be used to demonstrate our approach based on a scenario of operation to demonstrate the methodology's real-time capabilities. The train door system monitors and controls the process of the typical train door. This process concerns the safe boarding, departing and travel of train passengers. The system is comprised of a door controller, the door actuator, various sensors and the physical door itself. Figure 4 depicts the safety control structure of the system. The STPA analysis applied to this system identified 91 safety requirements in total. After removing the repeated specifications which were produced during the last step of the STPA analysis (i.e., from the translation of the unsafe control actions scenarios) the analysis defined the following: 1 accident, 3 hazards, 14 unsafe control actions, 4 improperly executed control actions and 36 safety scenarios (sample in Table 1). This system is used purely for demonstrative purposes and the complete analysis is not of much importance other than the example.

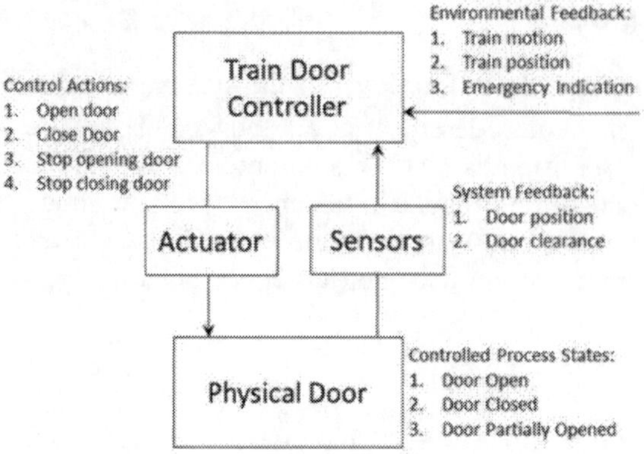

Figure. 4. Train Door System Safety Control Structure

Table 1. A subset of STPA Produced Safety Specifications of the Train Door System.

#	Symbols	Description	Connections
1	A	Passenger loss of life or injury.	-
2	H-1	The door is open when the train is moving fast.	A
3	H-2	A person is unable to leave the train (by the door) in case of emergency.	A
4	H-3	A passenger is caught by the door and the train is moving.	A
5	UCA5	Train Door Controller provides open door while the train is in motion.	H-1
6	CA3	Train Door Controller is unable to provide close door while the train is moving, and the door isn't closed.	H-1
7	S05	Small objects stuck in door movement rail.	CA1, CA3
8	S24	Evacuation mode misfire.	UCA5

4.1 Results

The scenario unfolds as follows. The train is stopped at the station and the door rail is filled with dirt, resulting in the door being unable to close. Then the train starts for the next station but while the command to close the door is issued by the door controller the actuator is unable to execute it, leaving the door open while the train is moving. A passenger then sees the door open and the train moving and scared presses the emergency button. This causes the system to change contexts of operation fundamentally changing the safety priorities of the system.

4.1.1 Model Development process

1. Creation of acyclic diagram. The safety constraints from the STPA analysis were translated into an acyclic diagram (Figure 5). The diagram is comprised of 143 distinct paths. In our example, only 2 of these paths are studied (Figure 6).

2. Time ranges. The time values for the connections of the safety constraints were arbitrarily assigned (sample in Table 2).
3. Selection of operation. Because in this system small changes can lead to big variations in the safety level the MINIMUM operation was used to discern the single time values (Table 2).

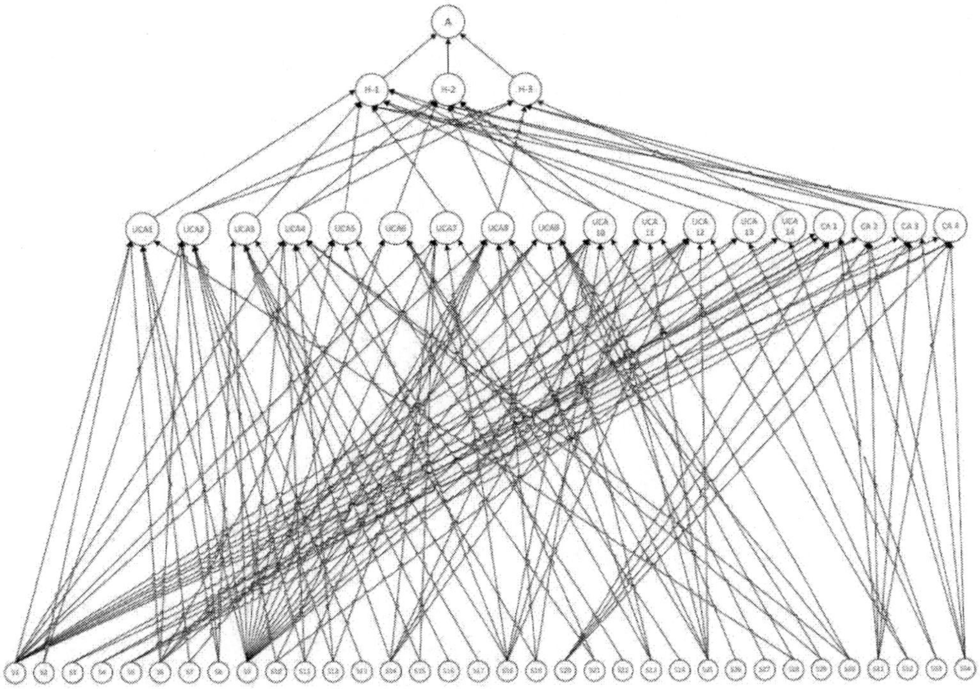

Figure 5. The complete acyclic diagram of the train door system.

Table 2. Time ranges and single time values.

Connection	Time range	Time value with minimum operation
H-1→A	[60,120]	MIN[60,120] = 60
CA3→H-1	[2,6]	MIN[2,6] = 2
UCA5→H-1	[10,30]	MIN[10,30] = 10
S05→CA3	[90,300]	MIN[90,300] = 90
S24→UCA5	[8,20]	MIN[8,20] = 8

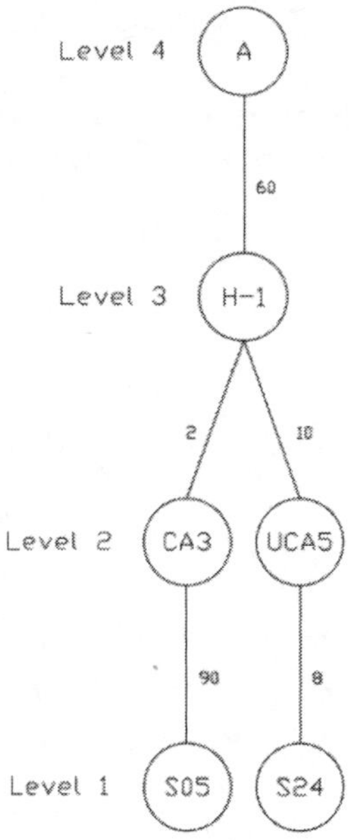

Figure 6. The acyclic diagram for the train door case study.

4.1.2 Safety Level Determination

1. Operational modes. The system is divided into two operational modes: the train stopped at a station, the train travelling between stations, but for the sake of simplicity in the example takes place only when the Travelling between stations Operational mode is active.
2. System monitoring data. The system specific data needed for the nodes of Figure 6 are: the door's state (open, closed), the train's velocity, doorway clearance, evacuation state, if the door movement rail is filled with small objects and evacuation mode misfire.
3. Mathematical model. In Figure 6 the sample of the acyclic diagram that will be used for the demonstration of the suggested approach is shown. This diagram contains 2 distinct system paths ($\vec{p}_1$: S05→CA3→H-1→A and $\vec{p}_2$: S24→UCA5→H-1→A). According to equations (3), (4) the values for the time remaining until accident for these two paths are the following:

$$t^{+}_{\vec{p}_1}(y) = \begin{cases} t^{\vec{p}_1}_{1\to2} + t^{\vec{p}_1}_{2\to3} + t^{\vec{p}_1}_{3\to4} = 90 + 2 + 60 = 152 & y \in \{0,1\} \\ t^{\vec{p}_1}_{2\to3} + t^{\vec{p}_1}_{3\to4} = 2 + 60 = 62 & y = 2 \\ t^{\vec{p}_1}_{3\to4} = 60 & y = 3 \\ 0 & y = 4 \end{cases} \tag{3}$$

$$t^{+}_{\vec{p}_2}(y) = \begin{cases} t^{\vec{p}_2}_{1\to2} + t^{\vec{p}_2}_{2\to3} + t^{\vec{p}_2}_{3\to4} = 8 + 10 + 60 = 78 & y \in \{0,1\} \\ t^{\vec{p}_2}_{2\to3} + t^{\vec{p}_2}_{3\to4} = 10 + 60 = 70 & y = 2 \\ t^{\vec{p}_2}_{3\to4} = 60 & y = 3 \\ 0 & y = 4 \end{cases} \tag{4}$$

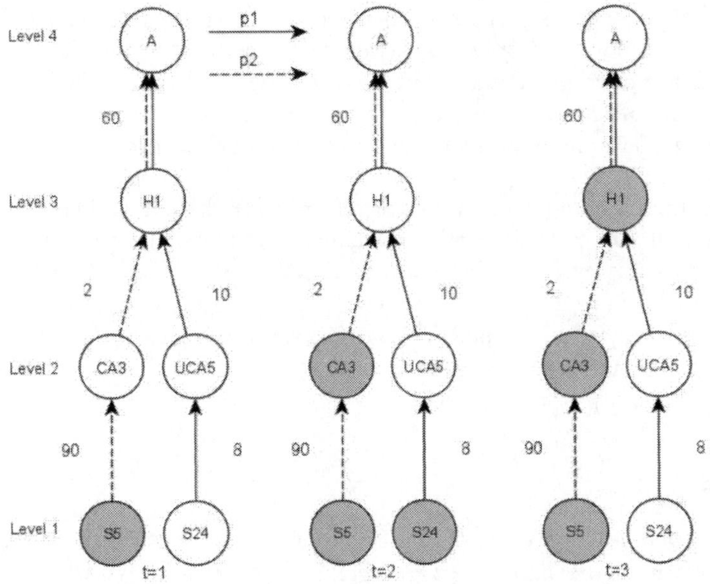

Figure 7. Example scenario of operation with acyclic diagrams for every time moment.

As Figure 7 shows, in the first time moment S05 (Table 1) is active this activates $\vec{\mathbf{p}}_1$ with values $\vec{\mathbf{p}}_1$= (1,152) and $\vec{\mathbf{p}}_2$= (0,78), so $\vec{\mathbf{p}}_w^{t=1}$=$\vec{\mathbf{p}}_2$ =(0,78), with the context of the system being that the train is stationary on the platform. In the second time moment the context changes to the train starting moving and gathering speed, with this change in context (that a computerized system can monitor with various sensors) the active nodes become S05 and CA3, S24 is active because a passenger has pressed the emergency button due to seeing the open door, these in turn activate $\vec{\mathbf{p}}_2$ with value $\vec{\mathbf{p}}_2$ =(1,78) and change $\vec{\mathbf{p}}_1$=(2,62), so $\vec{\mathbf{p}}_w^{t=2}$=$\vec{\mathbf{p}}_1$=(2,62). In the final time moment the train reaches cruising speed making H-1 active and in a few moments will start decelerating because of the emergency button being pressed this results in the two paths having the same value $\vec{\mathbf{p}}_1$= $\vec{\mathbf{p}}_2$ =(3,60), so $\vec{\mathbf{p}}_w^{t=3}$ = $\vec{\mathbf{p}}_1$=$\vec{\mathbf{p}}_2$ =(3,60).

5 Future Work and Concluding Remarks

By using this methodology, it is possible to identify the worst path to violate safety of a system with many possible paths that can lead to accidents. This can be useful to help safety managers and staff understand how system information and changes affects, safety and enable operators to react in timely manner to neutralize the completion of the path and hence the occurrence of an accident. The aim of this new approach is to enhance the capabilities of safety management systems towards preventing accidents and any types of losses by calculating in real time the safety levels of safety critical processes. This is feasible by projecting enriched information in a simple to understand form (second or minutes, instead of safety indexes) in real time as well as the acyclic diagrams, which can help managers understand how certain system interactions effect safety in their systems. This can be seen in the case study by observing that initially $\vec{p}_1$ seems more severe, if CA3 occurs the system gets much closer to the accident state from $\vec{p}_1$, and after the system reaches the point of Hazards the priority should be to stop the train and not meddle with the two different control systems S5→CA3, S24→UCA5.

It is the first time where a mathematical model is presented to address the problem of determining dynamically what the safety level is at a certain moment in time at a given context and calculate how much time is left for an accident to happen. This approach is also based on the results of an STPA analysis. This is another novel aspect of this paper since there are no previous concepts on real time safety level calculation based on the STAMP accident model.

A prominent limitation however of this novel approach is the fact that the mathematical model it is based on does not cope with uncertainty. It is planned, however, to address this problem in future work by enhancing the mathematical model with the use of fuzzy sets theory. An extension of the acyclic diagrams is also planned to take place in future work such that their acyclic nature would be nullified when system defences are in place. Finally, the creation of an operational system is planned where the approach will be installed into a real system to calculate its safety level in real time.

References

1. Øien K., (2010). Remote Operation in Environmentally Sensitive Areas; Development of Early Warning Indicators. *2nd iNTeg-Risk Conference, 14 – 18 June, Stuttgart, Germany.*
2. Thieme C. A., Utne I. B., (2017). Safety performance monitoring of autonomous marine systems. Reliability Engineering & System Safety. 159, March 2017, Pages 264-275.
3. Dekker S., (2006) *Resilience Engineering: Chronicling the Emergence of Confused Consensus*. Chapter 7 in Hollnagel, E., Woods, D. D. & Leveson, N. (Eds.), *Resilience Engineering: Concepts and Precepts*. Aldershot, UK: Ashgate, 2006.
4. Knegtering B., Pasman H., (2013). The safety barometer: How safe is my planttoday? Is instantaneously measuring safety level utopia or realizable?*Journal of Loss Prevention in the Process Industries*. **26**(4), 821-829.
5. Leveson N., (2015). A systems approach to risk management through leading safety indicators, *Reliability Engineering and System Safety.***136**, 17–34.
6. Karanikas, N. (2018). Documentation of Assumptions and System Vulnerability Monitoring: the Case of System Theoretic Process Analysis (STPA), *Proceedings of the 5th STAMP European Workshop*, 14-15 September 2017, Reykjavik University, Iceland, *International Journal of Safety Science*, 2(1), pp. 84-93
7. Chatzimichailidou M. M., Karanikas N., Dokas I. M., (2016). Measuring Safety Through the Distance Between Systems States with RiskSOAP Indicator, *Journal of Safety Studies*, 2 (2), 0-5.
8. Leveson N., Thomas J. P., (2018). STPA Handbook [on line], [Viewed 29November 2018] Available from: http://psas.scripts.mit.edu/home/get_file.php?name=STPA_handbook.pdf
9. Dokas I. M., J. Feehan and Imran S., (2013). EWaSAP: An Early Warning Identification Approach Based on a Systemic Hazard Analysis, *Safety Science*. **58**, pp 11-26
10. Leveson, N., (2011). Engineering a safer world: Systems thinking applied to safety. *MIT press.*

Safety differently: A case study in an Aviation Maintenance-Repair-Overhaul facility

Anastasios Plioutsias[1,*], *Konstantinos* Stamoulis [2], *Maria* Papanikou[2], and *Robert J.* de Boer[3]

[1]Coventry University, School of Mechanical, Aerospace and Automotive Engineering, Faculty of Engineering, Environment and Computing, UK
[2]Amsterdam University of Applied Sciences, Aviation Academy, Faculty of Technology, NL
[3]Northumbria University, Amsterdam Campus, NL

Abstract. You This paper presents the findings from a 'Safety Differently' (SD) case study in aviation, and specifically in a maintenance, repair and overhaul (MRO) organisation in Southeast Asia. The goal of the case study was to apply a new method of safety intervention that is part of the Safety Differently toolkit and utilises a bottom-up approach. This research tested the extent to which these interventions could be embedded into a continuous improvement program in a highly controlled environment, namely an Aviation MRO. The interventions (called micro-experiments, ME) are considered as a flexible tool, which allows testing of process improvements in a safe to fail way, empowering the lower levels of the organisation, challenging safety related issues and revealing key areas in need of transformation. The ideas for the interventions considered in the case study were retrieved from interviews conducted with 50 mechanics, and include issues to address aviation safety and occupational health as well as quality. We elected to include all three categories in this study as the ME approach is applicable to all of these. This MRO case study showcases the benefits and limitations of the ME in aviation, revealing the conditions under which it may become useful. Future studies should further explore the role of complex and heavily controlled industries in similar bottom up approaches, so that interventions can become part of a continuous improvement plan.

1 Introduction

Safety Differently is a new, yet under conceptualised, approach to improving safety processes by using the experience, knowledge and creativity of the frontline employees [1]. As part of the SD toolkit, micro-experiments (ME) have been introduced as a way to test the feasibility

[*] Corresponding author: ad3903@coventry.ac.uk

of interventions in a safe-to-fail manner. Maintaining aircraft to the highest safety standards requires skills of the mechanics within a Maintenance, Repair and Overhaul (MRO) organisation. Aircraft mechanics perform daily tasks, in demanding working conditions and whilst following procedures, designed to comply with safety and quality standards and regulations. Hence, given their subject-matter expertise, MRO personnel is a valuable source for the detection of issues experienced at the shop floor. Since it is important to continuously improve processes, this paper focuses on interventions utilising the most valuable source of an MRO organisation, its workforce. The MRO mechanics are key in sharing problems from the shop floor and help locate the potential interventions required to address issues affecting safety. Little is known about the results of the ME approaches and the added value to the continuous improvement of safety. Following the principles of Safety Differently (SD), we, therefore, explored bottom-up improvements through testing of the ME-based interventions. Our motivation stemmed from the lack of applications in highly controlled and complex risk industries, such as aviation. Aviation is an industry characterised by complexity, high severity risks, and consequently a strong regulatory framework. Any such ME case study or application in aviation is missing.

It was a challenge to examine the possibility of the ME application in an aviation organisation because the study must deal with strict regulations, whereas the methodology requires the elimination of constraints, possibly violating aviation rules. Our approach aimed to assess the benefits of ME in high risk industries and safety critical tasks [2] in two phases. The first phase involved data collection to a) identify gaps in the safety of the work processes and b) to design the subsequent potential interventions. The qualitative data was collected through a change-oriented research design, achieved by interviewing 50 mechanics who were the primary source according to the bottom-up approach of ME [3]. The second phase was the experimentation of a limited number of appropriate interventions, in line with previous applications of the concept outside aviation. The interventions were designed for their small-scale application by using the organisation's active work area and employees as an in-situ laboratory [4]. The project began in the base of the organisation with two university representatives who stayed in the MRO's base for months. The university agreed with the organisation about the project constraints, described in detail in 2.3, and the working progress of the project in the organisation's area and bays. In the following sections, we report the extent to which the SD-based methodology through the execution of ME could be used as part of a continuous improvement program within MRO organisations.

2 Safety Differently: concept and methodology

Sidney Dekker defines the Safety Differently (SD) concept as following: "Imagine a space, a space in which there are no rules, a space in which people can themselves determine the best course of action, a space in which people spontaneously negotiate and collaborate in order to create the safest outcomes for everyone, in this space a new type of humanity emerges, nobody is telling them, they figure it out on the spot, we call this emergent behaviour" [5]. In a complex environment, where behaviour is difficult to predict, we cannot be sure that the interventions that we devise actually bring the results that we envisage, or that unintended effects are avoided. We need an approach that allows us to try different things, does not cause peril, and can be stopped if we don't get the results that we want. This is where the micro-experiments (ME) come in.

In an MRO organisation, ME can be applied if the organisational safety constraints are removed in order to accommodate the bottom-up research design, examining how the mechanics have to maintain safety in their tasks and operations. Typical constraints in an MRO setting mean that wide-scale interventions are not suitable, mainly due to the number of stakeholders involved, such as airlines and regulatory bodies. An intervention can act as a safe-to-fail, small scale project, which uses the organisation's workplaces and workforce [6]. The subsections that follow discuss how the micro experiments are selected and how they were applied in the MRO organisation.

2.1 The ME concept

Captions Traditionally, aviation safety management is considered as a hard, positivist field of research [7]. Safety Differently involves taking a more collaborative approach to safety. It involves getting the people who do the work to decide on rules and safe working practices. This is different from the traditional approach to safety where rules and policies are set by managers and other third parties before being communicated to (or imposed on) the people doing the job. ME fit into this approach as they are safe-to-fail interventions that are monitored for success or failure, and that have been generated by the work-floor. ME are different to ad-hoc trial-and-error interventions in that they are planned before they are executed, and carefully monitored to see whether the intended results are realised. ME are different to projects in that projects are not supposed to fail, whereas in micro-experiments are temporary and we allow for unexpected, even disappointing results. In a complex environment we cannot have the (misplaced) confidence that any intervention will surely deliver the predefined results as expected. We need to allow for the possibility for the intervention not to work without causing undue damage. In fact the experiment can be considered a success if we realize an intervention "fails" and that some other solution is required. We therefore need to be able to stop or slow the experiments. On the other hand, if all goes well the intervention is ready to be formally implemented and possibly enlarged on a wider (geographic or organisational) scale. Naturally we will create the ideas for the micro-experiments in close collaboration with the task executors; many ideas will derive from them.

Micro-experiments generally require temporary safeguards because we will be deviating from existing processes, perhaps violating the constraints that are usually in place. Care needs to be taken to identify and only maintain those constraints that are absolutely vital to safeguard against unacceptable risks – we value the input of safety experts here. The criteria by which we determine whether an intervention is successful (or not) need to be determined ahead of time, although additional insights will occur as experience is gathered. Interventions that are deemed successful need to be incorporated in the next version of Work-as-Imagined. Even after formalisation, we remain vigilant about unintended effects, expecting the front line to report where exceptions are necessary. Multiple micro-experiments can run in parallel, preferably somewhat isolated from each other. Good targets for micro-experiments are procedures that are not delivering results as expected. Many organizations that we first engage with are susceptible to large and potentially embarrassing gaps between Work-as-Done and Work-as-Imagined, that are not straightforward to close. This is often a good starting point to try out micro-experiments. Other reasons to execute micro-experiments include adapting rules to changes in processes or equipment or for innovation. To further define the problem, we first need to identify the "job to be done": what operators really seeks to accomplish in the given circumstances. Successful micro-experiments help make the job safer and easier.

2.2 Methodology

Our methodology involved a sequential research design, comprising of two data collection phases. In the sections that follow, we initially describe the data collections phases for ME. Following this, we showcase the utilisation of previous findings in our data analysis. Before we present the results of the ME application, we describe the pre-coded themes and the analysis framework. Lastly, we discuss the limitations of these choices.

2.2.1 Research design phases

As outlined previously (cf. Section 1), the first phase involved the collection of data and the second phase regarded the interventions. The first phase included identifying and interviewing mechanics in semi-structured interviews. In order to identify the interviewees, the team requested to be present at the organisation's bays and observe the mechanics. Through this process, fifty (50) mechanics were purposively sampled and were interviewed individually. The main objective of this phase was to identify what gaps are prevalent between WAD and WAI. This data collection phase was supplemented with archival data from the MRO, including documentation from audits, investigations reports and safety reports, as well as injury reports and maintenance error notifications. Following this phase, the team identified a number of interventions, which formed the second phase of the study. The aim of the interventions as interventions was to close the gap(s) found through the findings from the first phase. During this phase, it was decided that the term interventions was changed to micro-projects following the organisation's requirement and in order to limit possible defensive behaviours that would affect results.

2.2.2 Coding and analysis framework

MRO organisations are typically optimised by lean approaches in order to create more value for their customers by utilising fewer resources [8]. Our approach hence initially required an assessment of the MRO environment. Before our study commenced, the quality (and safety) department at the MRO, had initiated a safety project to assess their safety performance. Elements of the latter are observed in the organisation's vision for the safety project itself, striving for innovation and evolution, and finally, Genchi Genbutsu, going down to the shop floor to fact check one's ideas in the 'real world' [8]. The findings of the project led to the establishment of a Safety Group (SG), encompassing the following pillars:

- Technical Systems: includes all the tools, equipment, and all the physical items in the organisation, necessary to create value for the customer.
- Management Infrastructure: includes the formal structures, the processes and the systems for controlling the resources, in association with the technical systems.
- Mindset and Behaviour: include the employees' thoughts and feelings as well as individual and group behaviour at the workplace.

These three pillars formed our initial classification that we used to compare the SD results with the organisation's SG project, and hence led to a pre-coding of the issues found at the shop floor. All the sources used from the organisation's safety project was provided to the team by the SG. The coded shops, which were chosen for phase one observations and consequent sampling and data collection (Table 1).

Table 1. Shop coding.

Shop	Code
Avionics shop	AVI
Structures shop	STR
Cabin shop	CAB
Cargo shop	CRG
Flight control shop	FLC
Landing gear shop	LDG
Engine shop	ENG
Seat shop	SS
Paint shop	PS

In the second phase, the researchers collected data, mainly from the interviews in which the front-line mechanics assigned. These technicians answered the questions posed to them and communicated their ideas for possible interventions. The findings from the interviews and observations led us to a master coding according to shop floor and their bays. In addition, the MRO's project used numerical values to score and classify findings following the Safety Risk Management (SRM) manual, which meets the compliance requirements according to ICAO's Safety Management Manual (SMM) and the data driven systems to proactively manage safety, Safety Management Systems (SMS). The MRO uses numerical values for the risk and taxonomies according to SMM standards [9], based on the Severity (minor to hazardous) and the Probability (extremely improbable to frequent). The Risk Indicator (R is calculated by multiplying the probability (P) with the Severity (S) of an occurrence (S), forming the $R = P \times S$ equation. The result of this equation rated from low to high risk and the values for the possibility and the severity explained detailed. The SRM was used during the project to score and classify the findings.

2.2.3 Research rigour

The team took a series of steps to minimise bias, gain access, and to overall ensure rigour in the research process. The mechanics were interviewed to gain information based on their expertise and bypass the organisational-level hurdles. By doing so, the feeling of empowerment of the mechanics increased, placing emphasis on the value of the technical and operational experience of the mechanics. In addition, we described the interventions to the organisation management to gain internal approval, allow execution of the interventions and to peer-review the results. Following this, and regarding the micro-projects, acceptance by management was ensured to reach consensus on the interventions proposed by the mechanics. After suitable interventions were collected, the gathered intel has to be verified by the rest of the mechanics, since implementing an idea that only suits one mechanic is not a valid micro-experiment. Besides validating with the mechanics, it is also important to validate the opportunities of improvement with the quality management representatives to make sure they agree on the findings and see the same promising interventions for current tasks or processes. In case there were multiple possible interventions, the mechanics scored them on an effort-effect diagram (Figure 1) together with the Quality Manager. All the proposed interventions were compared with each other in assuming the effort required to achieve each intervention and its expected effectiveness. An example is shown in Figure 1, where intervention "1" was preferred because it could yield the most effective result with the least amount of effort when compared to intervention "2". In general, interventions mapped on the top-right area of the matrix were preferred over the rest.

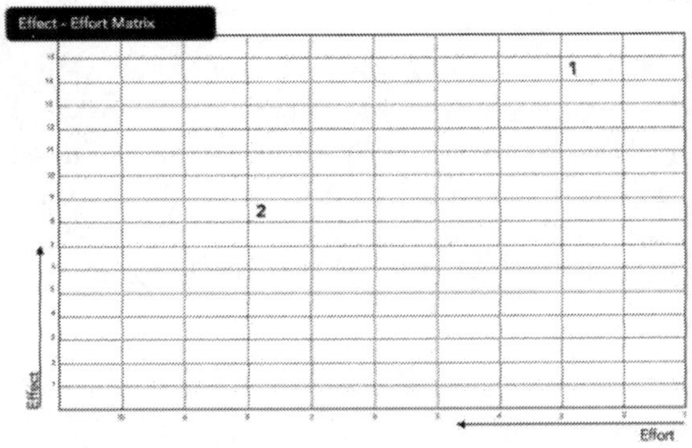

Figure. 1. Effect - Effort Matrix [10]

To gain further approval, the micro-experiment approach was described to management in detail, covering points such as the process, success criteria, constraints and anticipated interventions as well as the area and possibilities of amplification. Regarding the execution of the interventions, each micro-experiment had a customised period. The main factor was the duration of the task each intervention targeted. One researcher was continuously present during the execution of each experiment to observe the intervention, function as the point of contact for concerns, questions, ideas, and, most importantly, collect the views of the mechanics on each intervention. When the predetermined timeframe expired, the experiment was concluded by the researcher, the intervention was suspended, and the task was brought to its previous state (e.g. tools, procedures). A final template was created with the approval intervention when it validated from the quality management department. The decision on whether to amplify each successful intervention into an improvement plan laid with the organisation's management team.

2.3 Constraints

When the project was being planned, and following the principles of ME, the team observed that there are several limitations when applying ME in an aviation organisation. The chosen MRO organisation allowed the researchers time to organise and plan the project locally, but the quality manager and the safety group required the following constraints to be applied to the ME's:

- Any actions must be according to the regulations
- Any intervention must be according the procedures
- Any interaction cannot create new risks
- Any intervention is according to the organisation's risk assessment
- Any intervention is not a permanent solution
- Any intervention must comply with requirements set by organisation's clients
- Any intervention is only applied to the predetermined bay with predetermined group
- Any intervention will run for fixed period of time
- Any intervention can only be executed with the present of at least one of the university's representative on site.
- Any intervention can only be executed in agreement with Quality Manager, VP, the bay manager involved and the project team.

The limitations deriving from the application of ME in a highly controlled industry quickly became the first important finding of the case study. It is clear that the first three constraints (any actions must be according to the regulations, any intervention must be according the procedures, and any interaction cannot create new risks) do not allow any innovation to take place. These, as well as the required permission to execute the ME's, severely impeded progress at the site, and needed executive circumvention to allow the ME's to be applied. Although in the ME concept the interventions are described as a bottom up process, the validation, approval and decision processes in aviation are a top down process with assumptions and issues including not merely safety aspects but also extending to other departments like security and quality. The discussion about the limitations follows the presentation of our results.

3 Results

This section presents the findings from phases one and two of the sequential research design applied following the principles of the ME concept. Table 2 below provides an overview of the amount of data and of the sources involved in phase one of the study (i.e. interviews and archival data). Interview findings were initially categorised in single or multiple concerns. The categories emerged from the narratives of the research as presented in Table 2.

Table 2. Number of times that concern is raised and/or observed.

Concerns	Frequency	Number of findings
Single Concern	One time	18
Few Concerns	2 – 5 times	25
Multiple Concerns	6 – 10 times	8
A Lot of Concerns	>10 times	5

When a finding was a single concern, it meant one mechanic only raised it; such issues were either considered minimal, according to the QM and the SG, as general remarks or seen as irrelevant to the organisation. Valid concerns were findings reported by multiple mechanics, and issues on which the organisation agreed, but had its intervention planned, processed or implemented (11 in total). In addition, findings were categorised according to the bay coding (cf. Section 2.2.2) as presented in Table 3. As shown below, the findings were also categorised according to in- and on- the aircraft (in A/C, on A/C), and were summarised for each of the shops. Bay 6 is not included in the study.

Table 3. Numbers of findings per bay.

	In A/C				On A/C				
	CRG	CAB	STR	LDG	ENG	FLC	AVI	PAINT	Other
Bay 1	7	3	3			1	2		4
Bay 2		1	1						6
Bay 3		1							6
Bay 4		1	11						6
Bay 5		3	12	10			2		4
Bay 7									1
Bay 8				1				2	1

The shops that are not necessarily linked to all bays. For example, the landing gear shop (LDG), is linked only with the bay 5 and bay 8. In total, sixty-nine findings (69) were recorded. However, ten of these findings were finally selected as interventions, through the validation and approval process as described previously (cf. Section 2). The accepted findings as interventions are described in detail in 3.2.

3.1 WAD – WAI analysis

During the analysis of the interview data, it was observed that the mechanics were hiding the gaps between how they should do their job (WAI) versus the way they perform their tasks (WAD). Hence, we observed that the MRO organisation manifests signs of a problematic safety climate. These limitations are further discussed in Section 4.

3.1.1 Findings Characteristics

The findings initially examined and evaluated the possible interventions for the MRO organisation. As shown in Table 4, the intervention characteristics were classified using the three pillars of the SG, then by area of improvement and per type of method and lastly by using the MRO's risk assessment scaling method.

Table 4. Evaluation of Interventions

Classification of Findings		
1	Technical Systems	34
2	Management Infrastructure	21
3	Mindset and Behaviour	14
Type of method		
1	Process	50
2	WAD & WAI	15
3	Researcher's Observation	4
Risk Assessment of Findings		
1	High (>40)	3
2	Medium High (30-39)	8
3	Medium (7-29)	30
4	Low (<5)	28

In detail, in the risk assessment it is observed that eleven (11) findings are in medium-high and high risk and they have been rejected. The rest fifty-eight findings were examined and evaluated from the project team, the quality manager and the organisation's safety group. From the suggested area of improvement, the twenty four of the sixty-nine findings belong to safety area. The rest thirty-nine addressed to faciliatory and twenty-four are related with the equipment. Fifty of the total findings are related with the process, fifteen of them are with the WAD & WAI concept and four of them are related with the researcher's observations. When the team taxonomized with the same taxonomy of the previous safety group and QM project they found thirty-four (34) findings in technical systems (48% of the findings), twenty-one in management infrastructure (31% of the findings) and fourteen in mindset and behaviour category (21% of the findings).

3.2 The accepted interventions

Due to the absence of findings on safety alone, most of the processes discussed were not strictly safety-oriented but also overlapped with quality and occupational health domains. Although mechanics were hesitant in discussing these gaps, they were eager to suggest possible improvements and ideas. Similar to the problems of the WAD-WAI analysis, this process manifested that trust from the mechanics was low, and that, then, there is a problematic (safety) climate. During the process, and as the role of the researcher was distinguished from organisation management, trust increased, and the mechanics opened up, sharing the insights that follow in the ten (10) executed interventions, which have more quality characteristics than safety related issues.

Table 5. The accepted interventions

Problem description	Intervention description
The mechanics are not allowed to wear their safety shoes during final maintenance in a cleaned cabin, prepared for delivery to the customer. However, mechanics often make last-minute repairs in the cabin, in which the use of safety shoes is recommended.	Three types of covers were tested and a silicon, flexible shoe cover was selected due to ease of use, grip provided and anti-static properties. The experiment proved useful and QM extended the shoes throughout the organisation by adding them to the personal protection equipment (PPE).
Following installation the seat covers are often damaged and repairing them usually means delays in customer acceptance and is costly.	For the business class seat covers, the intervention is a fabric, reusable seat cover.
For small tools, such as a drill bit, a expendables form needs to be filled in at the tool room booth, making the tool keeper wait for the completion of the form.	The filling in of the form elsewhere can increase the throughput of the booth and therefore reduce the current queue time.
Under the wing of Airbus aircrafts, a strong presence of kerosene fumes can be noticed.	Fume-removing fans could be used to circulate the air. A risk, however, is that the fumes move to other work areas.
Current methods of working on the fuselage or wings of the aircraft can be dangerous to mechanics and can damage the aircraft.	*Butterfly Suction Cups:* A new system, in the shape of a butterfly, was proposed as an improved and safe alternative.
The safety shoes at are currently used by mechanics are uncomfortable and are of low quality.	The introduction of a new type/brand of safety shoes has been proposed.
The extreme hot climate working conditions in Southeast Asia.	To address the tropical working conditions, lighter shirts made from a dry-fit fabric, have been proposed for the mechanics.
When a tool is missing from one toolroom mechanics often walk to the other toolroom a few 100 meters away with time-consuming.	A transportation means such as a bicycle can significantly reduce the time needed to get a tool from a different tool room.
Everyday mechanics have to retrieve their standard toolboxes from the toolroom, which creates long queues at the start and finish of the shifts.	The Relocation of the toolboxes back in the shop area was suggested, and only using the toolroom as for original and limited equipment.
Inside the cabin, mechanics use stairs made from foam and tape, these are unstable.	A more firm and secure stair was proposed as a safer equipment for the mechanics.

4 Discussion and Recommendations

The SD principles guided the application of the Micro Experiments in a bottom up process to identify the problems and the solutions for safety related issues. The research objectives focused on exploring whether the application of the ME approach could trigger interventions for improved safety processes in an aviation organisation. Our two- phases methodology following the principles of the ME in an aviation organisation, revealed the benefits and the limitations of the approach in a high controlled and operationally risky environment. Below we discuss the key findings from the ME case study at the MRO organisation noting our contribution.

4.1 Occupational Health, Process Safety and Quality overlaps

One of the critical realisations during the project was the relationship between safety and quality management. Through the research, the team realised that the constraints and the limitations in an aviation organisation are strong barriers for the ME application as described and applied elsewhere. The lack of case studies in aviation meant that the research would enter an under-explored field, which created challenges and sometimes misunderstandings between the researchers and the management of the organisation. We therefore report that the ME assumptions are challenged by the organisation's quality and safety department, which descriptions of ME do not account for. For example, the MRO's quality department attempted to assess the safety findings and observations as quality markers. Our findings here highlight the problems in the quality vs safety and occupational health misperceptions, and how these create confusion in the safety management process and the identification of suitable interventions for the organisation. Secondly, the safety department initially raised constraints that paralysed the ME and needed to be lifted by top-level intervention for progress to be made

Albeit, the revelation shows that the ME approach can reveal deeper organisational issues intervening in the safety processes, whether managed or performed, and supports the benefits deriving from bottom-up approaches [3]. Because of these benefits, mid-project interventions were able to take place. Initially, the team was able to recognise the lack of ME fundamental knowledge and thinking in the MRO organisation. Secondly, the team was able to address this issue by filling the knowledge gap with ME workshops and additional visits on-site the MRO. The intermediate actions aided to have most of the interventions accepted from the organisation, yet, as reported in Section 3, with quality and occupational health aspects, and not a clear safety orientation. The same misconception was identified in the interviews where the participants wanted to discuss and proposed quality and not safety interventions like the cover of the cabin seats during the work of the technicians in the aircrafts' cabin. This finding and change process underpins the added value of ME and the intervention possibilities in aviation organisations.

4.2 Insights following the SD principles

Using the ME approach to inquiry, the findings from interviews in interventions achieved a variety of results and insights into the adoption of the boader SD principles. Concerns such as middle management isolation, fear of change, and not wanting to take responsibility, revealed problems of safety bureaucracy and showcased the possible issues these can induce during a (research) project and in the safety processes of an MRO organisation without directly impacting the ME. The mechanics also reported that there are even cases in which they were demoted after they made critical remarks about the organisation and/or shared innovative ideas. There are clear indicators of the presence of these types of bureaucracies within the organisation. Our ME findings hence support studies highlighting the importance of safety climate underpinning aviation MRO practice [11]. Importantly, our study showed that, when the organisation shows value and appreciation towards the work of the mechanics, their opinions and ideas, the attitude towards responsibility changed. Specifically, the MRO mechanics were appreciative of the responsibilities and chances are given to them to be part of a change-based project. It showed them the importance of bottom-up suggestions and challenging of organisation procedures. The interventions provided the flexibility required to fit and try a wide selection of interventions. The first completed interventions showed the amount of effort mechanics put in their ideas and the results they can achieve from this input. The fact that the shoe cover intervention (Table 5) will be amplified organisation wide is a first step towards the continuous improvement.

4.3 Evaluation of the MRO project

During the planning and execution phases of the project, several success factors and limitations were revealed. Success factors include the MRO's availability and willingness at a high level to run the project in their working area. The decisions the team took in the latter environment, aided in achieving research rigour and insights for the transferability of the results [12]. In detail, the team followed the data to make decisions instead of aiming to satisfy standards not relevant and not appropriate for the project (i.e. external and internal validity, reliability and objectivity). In addition, the researchers' professional background in aviation organisations helped establish a common communication channel with the MRO managers, gaining access into their MRO systems. Hence, familiarity with the culture of the participants aided reflexivity [13]. As a bottom-up process underpinned the project, the developing cooperation between managers and mechanics became evident. This success factor showed the power of the ME approach and the advantages over traditional intervention approaches (ad-hoc or projects).

Limitations of the project include the chosen context of the case study, hence addressing our main objective. In detail, in the highly controlled MRO environment, the more inexperienced members were not able to recognise possible safety-related issues, which would have been useful for the project process. Moreover, the inexperienced members unintentionally behaved like auditors, adding to a lack of trust already in the MRO, and revealing problems of safety climate. Similarly, the managers were unprepared for this type of project. Our findings hence support past aviation MRO safety studies, where safety climate is seminal to developments. Additional limitations were the context of safety bureaucracy, which surfaced organisational problems such as the lack of a positive safety culture, the lack of trust and the lack of safety knowledge in the middle and the low-level employees. For example, during interviews, it became evident that most employees are confused between the failure analysis and incident investigation.

5 Conclusion and future directions

Safety Differently in aviation organisations is an underexplored field. Micro experiments are not safety recommendations, but they comprise a tool to explore the perceptions of shop-floor employees that guide actions and behaviours. In order to formulate interventions, incident reports can be used, although it should be noted that incident reports do not trigger them. Incident reports can be used as a tool to identify focus areas, assess the status of the safety recommendations and close the loop. In addition, it is recommended that, although the micro experiment is a bottom-up process, the organisation is assessed before any safety differently project can begin. Our case study application revealed the bottle neck situation which forms our contribution to further ME applications in a similar aviation organisation. Our recommendation is hence to initially assess the organisation's safety climate and then to apply the framework for the Safety differently project. The assessment of safety climate is fundamental for the SD – and therefore ME - implementation in aviation. It is a way to plan the researcher's actions before the application of any method and tool of SD concept. The preparation of the safety program participants is an additional necessary action, which includes top-down training and identification of the setting the micro experiment takes place, so that daily working load is not affected. They are no potential risks for the organisation's operations. Moreover, it is recommended that the organisation prioritises the organisation's problems, using quality tools and methods (e.g. Effect – Effort, Pareto, Fishbone diagrams).

In any aviation organisation which has many working tasks, the term 'micro experiments' is not a familiar one and could, therefore, threaten their acceptance. It is therefore suggested that the term changes to 'micro-projects interventions', which is a less threatening term to aviators. Future research should hence focus on tactics for the smoother application of the micro-project interventions can give opportunities in organisations to prepare, plan and organise better large-scale projects in the future. With the use of interviews at the shop floor, new findings and interventions can be collected, executed and possibly amplified throughout the organisation. However, mechanics will have to be empowered to share their concerns and ideas openly. Addressing safety bureaucracy problems will not only transform the intervention to an improvement plan, but it can help enhance wider safety practices (e.g. safety promotion, reporting). Safety Differently and ME can, therefore, be used to continuously improve following a bottom-up approach.

Acknowledgements

The authors would like to thank Juno Beckers and Niek Kuilder, for their contributions.

References

1. Dekker, S. (2015). Safety, Differently. Boca Raton: CRC Press.
2. Weber, D. E., & Dekker, S. W. (2017). Assessing the sharp end: reflections on pilot performance assessment in the light of Safety Differently. Theoretical issues in ergonomics science, 18(1), 59-78.
3. Hale, A., Borys, D., & Else, D. (2016). Management of safety rules and procedures: a review of the literature. Report submitted to the IOSH Research Committee, Report 12.3.
4. Gantt, R. (2018). Safety Differently- A new view of safety excellence. Available online at https://safetydifferently.com/wp-content/uploads/2018/10/Safety-Differently-ASSE-Proceedings-Paper.pdf. In Dekker, S. (2018). A micro-experiment. In S. Dekker, *The Safety Anarchist* (pp. 180- 185). Oxford: Routledge.
5. Dekker, S. W. A. (2017). Safety Differently, The Movie. Available online at https://safetydifferently.com/safety-differently-the-movie/.
6. Dekker, S. (2018). A micro-experiment. In S. Dekker, *The Safety Anarchist.* Oxford: Routledge.
7. Ferroff, C. V., Mavin, T. J., Bates, B. R., & Murray, P. S. (2012). A case for social constructionism in aviation safety and human performance research. Aeronautica, 2(1), pp. 1-12.
8. Nanova, G., Dimitrov, L., & Neshkov, T. (2012). Lean Manufacturing Approach in Aircraft Maintenance Repair and Overhaul. Sofia: Technical University of Sofia, Bulgaria.

9. ICAO (2013). Safety Management Manual. Available online at https://www.icao.int/safety/fsix/Library/DOC_9859_FULL_EN.pdf.
10. Beckers, J. (2019). *Process Improvements with Safety Differently*. Amsterdam University of Applied Sciences.
11. Fogarty, G. J. & Shaw, A. (2010). Safety climate and the theory of planned behaviour: towards the prediction of unsafe behaviour. Accident Prevention and Analysis, 40(5), pp. 1455-1459.
12. Guba, E. G. & Lincoln, Y. S. (1981). *Effective evaluation: Improving the usefulness of evaluation results through responsive and naturalistic approaches*. Jossey-Bass.
13. Shenton, A. K. (2004) "Strategies for ensuring trustworthiness in qualitative research projects", Education for information, 22(2), pp. 63-75.

Neuroscientific tools in the cockpit: towards a meaningful decision support system for fatigue risk management

Maria C. Papanikou[1,*], *Christos* A. Frantzidis[2,3], *Anna* Nikolaidou[2,4], *Christina* Plomariti[2], *Maria* Karagianni[2], *Vasilis* Nigdelis[2], *Aliki* Karkala[3,4], *Christiane* Nday[2], *Giorgos* Ntakakis[2], *Akis* Krachtis[2,4], *Panagiotis* D. Bamidis[2,3], and *Chrysoula* Kourtidou -Papadeli[2,3,4]
[1]Amsterdam University of Applied Sciences, Faculty of Technology, Aviation Academy, The Netherlands
[2]Aristotle University of Thessaloniki, Medical School, Laboratory of Medical Physics, Greece
[3]Greek Aerospace Medical Association-Space Research, GASMA-SR, Greece
[4] AeroMedical Center of Northern Greece, Greece

Abstract. Fatigued pilots are prone to experience cognitive disorders that degrade their performance and adherence to high safety standards. In light of the current challenging context in aviation, we report the early phase of our ongoing project on the re-evaluation of human factors research for flight crew. Our motivation stems from the need for aviation organisations to develop decision support systems for operational aviation settings, able to feed-in in the organisations' fatigue risk management efforts. Key criteria to this end are the need for the least possible intrusiveness and the added information value for a safety system. Departing from the problems in compliance-focused fatigue risk management and the intrusive nature of clinical studies, we report a neuroscientific methodology able to yield markers that can be easily integrated in a decision support system at the operational level. Reporting the preliminary phase of our live project, we evaluate the tools suitable for the development of a system that tracks subtle pilot states, such as drowsiness and micro-sleep episodes.

1 Introduction

Fatigue, stress and other conditions (e.g. mild stroke) can lead to micro-sleep episodes and subtle pilot incapacitation, meaning the progressive deterioration of a pilot's state, which "escapes the normal pilot medical screening" [1,2]. The deterministic approach to cockpit technology, the innovation research exploring single-pilot operations, as well as the current crisis in the aviation industry, form a context in commotion. In the covid-19 crisis, there is an unravelling discussion on replacing passenger flights with cargo operations, implicating longer flights, and, combined with night work, a more demanding and disruptive work environment. In addition, many pilots have lost their jobs, inducing greater distress,

* Corresponding author: m.papanikou@hva.nl

subjective affective states, manifested also in symptoms in the central and peripheral nervous system of the "survivors" [3]. These conditions are likely to impact sleep quality in terms of inability to fall asleep (prolonged sleep onset), frequent awakenings and non-invigorating sleep associated with attenuation of sleep depth (shorter slow wave activity duration). It is therefore argued that, a lack of human performance data system, allows lurking fatigue risks such as drowsiness to go undetected, and alertness levels to decrease during the critical flight phases (e.g. approach and landing). In addition, latest research highlights the need for airline level interventions [4] addressing pilot fatigue detection and including mental health. Previous studies, however, mainly propose models unable to capture complex human factors, due to isolated data collection, intrusiveness of detection and cost limitations. Aiming to enhance the arsenal of the neuroscientific tools, we propose a neuroergonomics framework, which consists of a detailed polysomnographic (PSG) sleep study accompanied with an elaborate neurocognitive and neurophysiological assessment. The proposed framework aims to elicit risk factors that are associated with fatigue patterns and to provide the appropriate mitigation approaches through non-pharmacological interventions. Developments regarding machine learning recommendation systems and potential intervention approaches are also described.

2 Fatigue risk management

At the operational level, in an effort to proactively manage aviation safety, safety management systems (SMS) have been set up to drive decisions based on data. Fatigue (risk) management involves the development and execution of a data-driven Fatigue Risk Management System (FRMS). Due to concrete limitations towards a truly data driven support system, fatigue is addressed as an isolated rest problem, and crew scheduling follows aeromedical requirements, the Flight Time Limitations (FTLs) to proactively manage fatigue. Aeromedical regulations and the selection and assessment of pilots do not predict future performance and health conditions but rather evaluate the condition of pilots at the time of assessment. Fatigue, however, may also induce boredom or even sleepiness which further deteriorates the pilots' psychomotor performance. Specifically, cockpit design and automation are linked to pilot boredom, and concerns of complacency and distraction are raised in the aviation community [5]. In addition, recent studies [4] note that the pilots are concerned about losing their licence, with 70% of the pilots being reluctant to report fatigue and wellbeing problems. Fatigue risk management hence requires the appropriate decision support systems (DSS) incorporating pilot characteristics, yet with operationalisable solutions.

In light of the latter gaps and the aforementioned focus on compliance, detection and pilot referral is lacking development at an operational level and fatigue management is based on regulatory guidance (i.e. FTLs). As a result, pilot fatigue interventions traditionally focus on self-management techniques. For example, studies [6] report that the top three self-management mitigation measures for fatigue include: a) more sleep, b) reducing workload, and c) being self-aware and making healthy lifestyle choices. Accordingly, fatigue studies are based on FTLs and fatigue risk prediction for better crew planning using mathematical models [7], aiding compliance in operational planning. From these and similar approaches tools were developed, such as the Boeing Alertness Model (BAM) that focuses on capturing sleep inputs in order to aid planning and scheduling. The limitations of BAM are highlighted by the Flight Safety Foundation, where they note that the model lacks validation within an airline environment. Other efforts to monitor fatigue include applications such as the Jeppesen Crew Alert, aiming to complement FRMS. The use of such tools is intended for pilots, who can insert personalised settings in order to record sleep patterns, and to generate and share fatigue reports. An advantage of this approach is that it helps pilots to be more aware of their alertness levels. Such tools extend to assessing controlled rest. For example, an actigraphy watch is used for the detection of sleep disorders. Latest applications in research show the positive effects of the latter's use for controlled rest [8], yet, despite individual pilots reporting the benefits of its use, applications at an airline level are not reported.

Neuroscience offers more insight into fatigue management by deploying sleep studies, which, however, are not utilised for operational improvements but mainly highlight the importance of fatigue and criticise the FTLs, engaging in policy making discussions. Eye-tracking studies are also popular in fatigue detection [9], as well as the use of additional tools, such as the Psychomotor Vigilance Task (PVT), to assess reaction times and attention. Latest applications of the PVT include short test flights (approx. 30 minutes each) in simulated environments [10]. Importantly, however, a recent study of pilot cognitive workload using headbands in real – time operations showed differences from a simulated environment [11]. Significant differences between simulated and real-time environments are reported, where pilots commit errors they did not in the simulator [11]. Specifically, the study reports that in high workload phases of flight, errors were committed in the simulated environment were less that in the real-time environment. Hence, although, the majority of studies report lurking risks and the benefits of risk managing fatigue beyond the FTLs, there is no operational-friendly system that is able to create, simple yet informative, data banks the FRMS requires. The papers on the value of biomedical engineering and data mining far exceed the acceptance of their proposed solutions for flight crew performance.

In addition, sampling limitations are noted as previous neuroscience-focused studies included student [12] or inactive pilots [13]. Similarly, pilot characteristics and performance, and their challenges in complex environments, do not go beyond psychological profiling for selection purposes. In light, however, of studies documenting the pilot's professional changes [14, 15], we propose that the inclusion of professional characteristics when studying subjects can help develop a better decision system. Decrease of training, safety practices led by compliance and cost-cutting needs, dependency on employers, the imbalance of investments between the human and the machine (human out of the loop), lack of reporting, low status and discretion were found as the long-term changes induced to flight crew as the industry grew and changed [14]. Airmanship 2.0 [15] deals with the skills required to operate an aircraft, extending from manual skills and the level of automation (LoA) to macro-cognitive factors and key areas of concern. In particular, pilots are found re-skilled in terms of gaining new flying skills due to automation and de-skilled in terms of manual flying. Key areas of concern include fatigue management, failure diagnostics, emotional self-control, complexity oversight and knowledge of manuals. For the purpose of our project, we distinguish between *Automation* and *Employment*. The latter encompasses elements that can amplify the areas of concern under Airmanship 2.0. Based on the latter studies, Table 1 summarises the framework that underpins our live study. We distinguish between the Old and New generation of pilots and, automation and employment characteristics. Age is included due to its significant relation with cognitive abilities and with knowledge, meaning the expertise, acquired skills and procedural knowledge [16].

Table 1. Airmanship 2.0 adapted from [14,15]

	Airmanship 2.0	
	Automation	Employment
OldGen 46-65	De-skilled (Aviate-navigate-communicate-manage)	Reduced training-decay of knowledge- less discretion

	Airmanship 2.0	
	Automation	Employment
NewGen 20-45	Re-skilled (Manage-communicate-navigate-aviate)	Debt of training-low income-employment dependency

We return to our underpinning framework in light of the preliminary findings reported in this paper. Before engaging with the data, we describe the deployed methodology for this early phase of our project using three cases, one from the OldGen typology and two from the NewGen typology.

3 Methodology

In this section we describe the underlying data acquisition as well as the analysis employed for identifying potential, fatigue risk factors. The methodological objectives support our key aim, namely developing a DSS that is enabled by contextual characteristics. We deploy three cases in order to explain our methodology. Each case underwent a different set of experiments and assessment in order to explore intrusiveness levels and operational value. The latter aimed in a) determining the value of tools for an operational setting, and b) exploring the value of pilot profiling in the interface with the tools. Below we explain the neuroscientific tools for the experiments, and we describe the cases that represent the typologies of OldGen and NewGen pilots.

3.1. Polysomnographic data acquisition

A Nihon Kohden electroencephalographic (EEG) device equipped with 32 Ag/AgCl electrodes was used. The EEG data were collected from 19 electrodes located according to the 10-20 International System. A ground electrode was placed on the prefrontal midline (Fpz) positions and two reference electrodes were placed on the left and right mastoids. The electrooculogrammic (EOG) activity gathered both vertical (blink) and horizontal movements by placing bipolar electrodes above and below the right eye and on the outer canthi of both eyes. Electromyographic (EMG) activity was recorded by a bipolar electrode placed on the left and right chin, whereas electrocardiographic (ECG) timeseries was recorded by a bipolar electrode placed above the chest.

Pre-processing pipeline

The entire night PSG data were initially divided into 10 segments of equal size in order to facilitate more efficient computations. From each segment a baseline removal took place initially by subtracting from each signal its mean value. This resulted in eliminating amplitude linear shifts (DC bias). Regarding the EEG data a common average re-reference procedure was performed. Then, digital filtering was performed through Butterworth filters of 2nd order as follows:

1. High-pass filter (cut-off frequency 0.5 Hz) for further removing linear trends
2. Low-pass filter (cut-off frequency 50 Hz) for rejecting unnecessary high frequency content
3. Band stop (notch) filter among 47-53 Hz for removing industrial noise centered at 50 Hz
4. Band stop (notch) filter among 97-103 Hz for removing industrial noise harmonics centered at 100 Hz

5. Band stop (notch) filter among 147-153 Hz for removing industrial noise harmonics centered at 150 Hz

In the case of the ECG signal the cut-off frequency of the low-pass filter was set at 20 Hz and at 15 Hz for the EOG. In the case of the EMG signal the cut-off frequency of the high-pass filter was set at 10 Hz and for the low-pass at 70 Hz. Then, the algorithm of the Independent Component Analysis (ICA) was ran for each segment through the EEGLAB user interface under Matlab environment. Visual inspection of the ICA components by two experienced neuroscientists resulted in the rejection of artefactual sources due to blinks, muscle movements, bad electrode placement, ECG modulation. Finally, the data were epoched into continuous, non-overlapped epochs of 30 second duration and two experienced sleep experts were manually scored according to the guidelines of the American Association of Sleep Medicine (AASM) [17].

3.2. Cognitive screening

Neurocognitive assessment was performed through a single session of the CogScreen aeromedical edition (Cogscreen AM). The examination was performed in the premises of the AeroMedical Center and was performed by two experienced medical doctors. It is a computerized cognitive battery able to detect even subtle alterations in cognitive functioning. Although CogScreen does not focus on aviation knowledge or flying skills, it could provide quantified estimations of the underlying perceptual cognitive, and information processing abilities associated with flying. An initial rough estimate is provided by the Logistic Regression Probability Variation (LRPV) score of brain dysfunction. This ranges from 0 to 1 and the higher the score the more likely a cognitive degradation is. However, it is highly correlated with physiological aging and its value represents only the probability of impairment existence and not its impact. Further information is provided by a base rate analysis, which describes the number of tasks that the participant scored lower than the 5th percentile and 15th percentile when compared to a selected normative comparison group. This group mainly consists of astronauts, US military and civil aviation pilots. However, it is possible for the examiner to establish her/his own database of pilots [18]. There are also tasks associated with aviation factors such as deductive reasoning, motor coordination under speeded conditions, visual learning and recall, visual scanning, perceptual speed, and working memory and visual/psychomotor tracking accuracy.

3.3. Cases

In this report, we include three cases of male Caucasian right-handed active pilots. The pilots were assessed between flights following their roster. The sleep studies took place in the night between their flights and the other tools were utilised in their day off. The participants did not report any medical or psychiatric history until now. The participants did not suffer from any neurologic or psychiatric disorder or human immunodeficiency virus (HIV). More specifically, the participants' medical history did not contain any instance of hypertension, kidney disease, diabetes, liver, heart, respiratory disease or cancer diagnosis. Neurological examinations did not reveal any head injury, stroke, brain tumor or seizure. Psychiatric examinations did not reveal any anxiety, depression, learning disabilities, mood or attention deficit disorder. Furthermore, the participants were systemically sustained from alcohol or any drug substance consumption. Different characteristics were taken into account in their selection, such as aircraft type, rank, airline business model/operations, and career progression.

OldGen
The first profile was of an airline captain (Case 1). Case 1 was aged 53 years old, with 18 years of education (basic education plus six academic years). He was an active captain for a large full-cost carrier with an ATP license type and 18000 total flight hours. Following the sleep study and the Cogscreen AM session, the participant was interviewed, in order to assess his perception of the current challenges, and to assess acceptability, intrusiveness feelings, and to compare feelings and perceptions with the experiments' results. The interview lasted 1hour and 28 minutes. The semi-structured interview protocol followed previously validated factors in order to develop the coding framework for analysis based on a) mental and temporal demands, b) perceived performance, and c) frustration fatigue. Physical workload was excluded, whereas temporal workload was analysed based on time pressure and pace of tasks, due to the different phases of flight.

NewGen
The second profile was of a first officer (Case 2). Case 2 was 35 years old, with 16 years of education (basic education plus four academic years). He was an active ATP licensed first officer for a small full-cost airline operating a Bae 146 with 4000 total flight hours. Case 2's state was assessed only through the sleep diagnostic examination. Finally, the third profile was of a male first officer (Case 3). Case 3, aged 32 years old, with the same level of education as Case 2. He was an active ATP licensed first officer for a small low-cost carrier operating an A320 with a total of 2000 flight hours and was at the time preparing to transition to the captain's role. Case 3 was interviewed following the same protocol as with Case 1. The interview lasted 2 hours and 15 minutes. Case 3 was excluded from the sleep study but was included in the Cogscreen AM session.

3.4. Neurophysiological feature extraction for drowsiness estimation

The estimation employs features derived from the ECG, EMG and EOG in order to discriminate active awake from drowsiness. The ECG analysis involves time-frequency analysis of the heart rate variability (HRV) through the Lomb-Scargle periodogram (LSP) [13]. Firstly, the ECG signal was normalized and then R peaks were detected, forming thus the R-R timeseries. Since, the R-R timeseries does not follow a uniform sampling like the other biosignals we extracted its spectral component employing more advanced mathematical tools than the Fourier transform. More specifically, the LSP method resulted in the calculation of the following spectral features: The frequency with the maximum HRV power and the relative energy ratios of the ultra-low Frequency (<= 0.0033 Hz), very low frequency (0.0033 – 0.04 Hz), low frequency (0.04 – 0.15 Hz), high frequency (0.15-0.4 Hz), very high frequency (>= 0.4 Hz) and the range in which the frequency with the maximum power was detected. The EOG activity was estimated as 1) the number of blinks per minute (blink rate), 2) mean blink amplitude, 3) the mean rise time and 4) blink duration. The EMG spectral features were estimated as the relative energy contribution of very low (<=4 Hz), low (4-8 Hz), medium (8-12 Hz), high (12-24 Hz), very high (24-48 Hz) and ultra-high (48-96 Hz) of the chin EMG activity. The Orthogonal Discrete Wavelet Transform (ODWT) was employed for estimating the energy contributions with optimal time-frequency resolution [14]. The analysis involved 18 instances of active wake and 23 instances of drowsiness. The identification of each instance was assessed by two sleep experts. Each instance duration was 1 minute (60 seconds). In summary, the analysis involved 41 minutes of PSG recordings.

4 Findings

4.1. Case 1

The coding framework (cf. 3.3) was applied on the interview data. As shown in Table 1, the participant shared concerns about airmanship, characterising the cockpit "so virtual [that] it is like a video game", without having the knowledge of what lays "behind the system". Beyond flying skills, the decision making, and other macro-cognitive and resource management factors are affected. The participant's perceived mental workload was high in terms of demands in effort and tasks that do not only include the cockpit ones but 24/7 engagement with the company, online trainings, and dual roles. In detail, the participant reported an "extremely high" workload, and working in an environment where "everybody is tired" and employees are "24/7 [..] company-minded". Temporal demands were discussed in different flight phases with an emphasis on alertness in critical phases. Furthermore, frustration fatigue was discussed through the profession's changes affecting their knowledge, skills, and abilities (KSA). Employment concerns amplified frustration fatigue, involving emotions of stress, anger, and cynicism. In this context, the participant stressed the importance of experience to mediate lurking risks. The participant also noted the increase in his flight hours and his reluctance to report in the beginning of his career, whilst new pilots were characterised as ones lacking stimuli. The participant reported the intention to accept a wearable device in the cockpit.

Table 2. Perceptions Case 1.

Mental demands	Temporal demands	Perceived performance	Frustration
Effort: high	Time pressure: extremely high	Accomplished tasks: Manual	Profession: deterioration of KSA
Tasks: continue outside the cockpit.	Pace of tasks: irregular, lack of stimuli at cruising altitude.	Least satisfaction: Decision making Information processing Reaction time Team leadership	Employment: security, aggressive environment.

Sleep macro-architecture

The participant slept for 244.5 minutes (Figure 1). The sleep latency recorded as the duration from the first epoch with eyes closed and drowsiness until the first sleep (NREM1) epoch and was at 23.5 minutes. The duration of each sleep stage was 58.5 minutes for N1, 148.5 minutes for N2, 18 minutes for N3, 19.5 minutes for REM. Latency periods were 37 minutes for N2, 138 for N3 and 252 for REM. The slow wave activity (SWA) covered the 7.36% of the total sleep time, while the REM period covered the 7.98% of the total sleep time. Moreover, the participant woke up 11 times and the total awakening period was 77.5 minutes. So, its sleep efficiency measured as the total sleep time divided by the entire PSG duration was estimated at the 65.2%. Finally, the participant experienced 44 arousal episodes. There were also 55 spindles and 105 K-complexes.

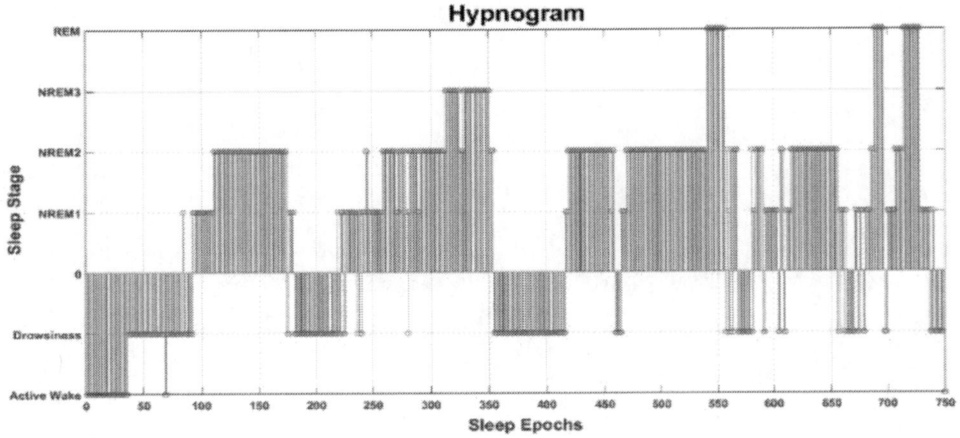

Figure. 1. Hypnogram of OldGenCap representing the sequence of the sleep stages across the entire sleep recording through polysomnography

Neurocognitive Assessment

The LRPV score of Case 1 was 1. The aviation factor scores are displayed in Table 3.

Table 3. Case1 performance on cognitive domains associated with aviation characteristics

Factors	Description	z-score	T-score
Attribute Identification	Deductive reasoning	-4.14	8.56
Motor Coordination	Motor coordination under speeded conditions	-0.57	55.70
Visual Association Memory	Visual learning & recall	-2.08	29.19
Speed/Working Memory	Visual scanning, perceptual speed & working memory	-0.32	46.84
Tracking	Visual/psychomotor tracking accuracy	0.10	48.97

4.2. Case 2

Sleep macro-architecture
Case 2 slept for 471.5 minutes (Figure 2). The sleep latency was at 24 minutes. The duration of each sleep stage was 172 minutes for N1, 212.5 minutes for N2, 77 minutes for N3, 10 minutes for REM. Latency periods were 30 minutes for N2, 44 for N3 and 88.5 for REM. The slow wave activity (SWA) covered the 16.33% of the total sleep time, while the REM period covered the 2.12% of the total sleep time. Moreover, the participant woke up 18 times and the total awakening period was 53 minutes. So, its sleep was estimated at the 82 %. Finally, the participant experienced 18 arousal episodes. There were also 27 spindles and 111 K-complexes.

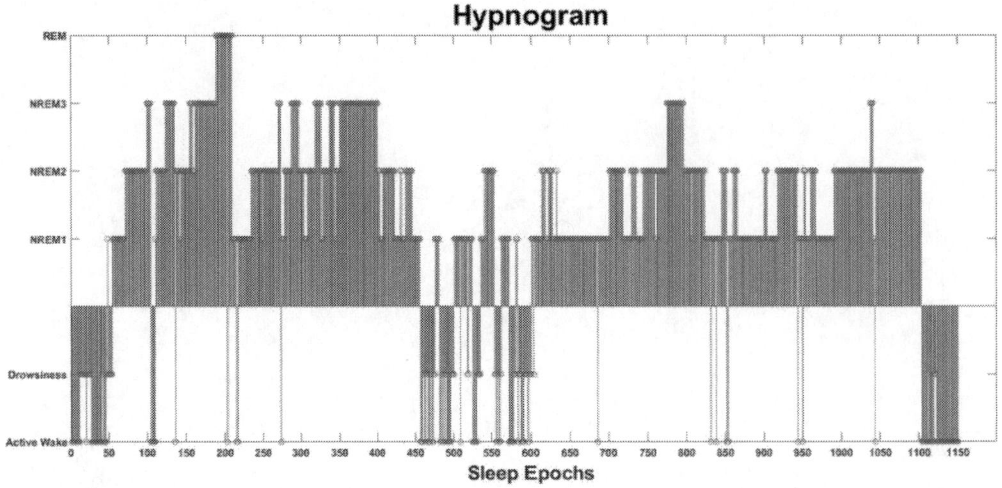

Figure. 2. Hypnogram of Case 2 representing the sequence of the sleep stages across the entire sleep recording through polysomnography

4.3. Case 3

Neurocognitive Assessment

The LRPV score of Case 3 was 1. The aviation factor scores are displayed in Table 4 below.

Table 4. Case 3 performance on cognitive domains associated with aviation characteristics.

Factors	Description	z-score	T-score
Attribute Identification	Deductive reasoning	-2.33	26.70
Motor Coordination	Motor coordination under speeded conditions	-2.03	70.28
Visual Association Memory	Visual learning & recall	-0.22	47.85
Speed/Working Memory	Visual scanning, perceptual speed & working memory	-1.57	34.30
Tracking	Visual/psychomotor tracking accuracy	-1.16	61.58

Case 3 interview data are presented in Table 5. Case 3 was frustrated and discussed in depth his concerns about the use of FTLs in small airlines, where rest according to regulations is followed to risk manage fatigue. Case 3 focused on the operational needs of airlines indicating that "what can be done in the minimum will be done in the minimum" and showing the interviewer a heavy night shift roster. Case 3 reports signs of accumulative fatigue, stating that he is flying the maximum hours possible and then returning to work tired "even after days off". "Bad management of flight crew" and that FRMS "is not in place", centered the discussion showing lack of peer or company support. Daily incidents in the airline's operations were characterised as "safe failures", but several AOGs (aircraft on ground) put more pressure on the operations. Regarding the temporal demands, Case 3 noted that there is a decrease in workload but that it is critically high in take-off and approach. Job insecurity (frustration fatigue) and high mental demands were noted as by Case 1. However, tasks are perceived as routine, and performance is perceived as appropriate for operating the aircraft. As in Case 1, Case 3 reported the intention to accept a wearable device in the cockpit.

Table 5. Perceptions Case 3.

Mental demands	Temporal demands	Perceived performance	Frustration
Effort: high	Time pressure: high	Accomplished tasks: Operating the aircraft	Profession: skills vs. knowledge of systems
Tasks: routine	Pace of tasks: high demand in critical phases of flight.	Least satisfaction: Developmental training	Employment: insecurity

4.4. Neurophysiological drowsiness estimation

The Matlab software was used to develop a drowsiness classifier based on the aforementioned features. We explored decision trees (fitted binary classification decision trees) as potential classifiers in order to result in a rule-based algorithm that would be easily interpreted in aviation and industrial settings. More specifically, we used active wake and drowsiness data from the two participants το train and test the classifier. There were 32 active wake and 33 drowsiness data epochs. The train set consists of 44 epochs and the test set of 21 epochs. The decision tree in Figure 3 was developed based on the train data and then its performance was validated in the test data, which were not previously used in the training phase. The root (most important feature) of the decision tree was the low frequency of the HRV timeseries. Then EMG low frequency and blink rate (number of blinks per minute) were demonstrated to further discriminate among the two classes. Finally, the blink amplitude was also an important factor. The resulted tree correctly classified 17/21 of the test instances. So, its classification performance was at 80.95%. A more detailed description of the method's classification accuracy is the following: a) True Positive 9/10 (90%), b) True Negative: 8/11 (72.73%), c) False Positive: 3/11 (27.27%), d) False Negative 1:10 (10%).

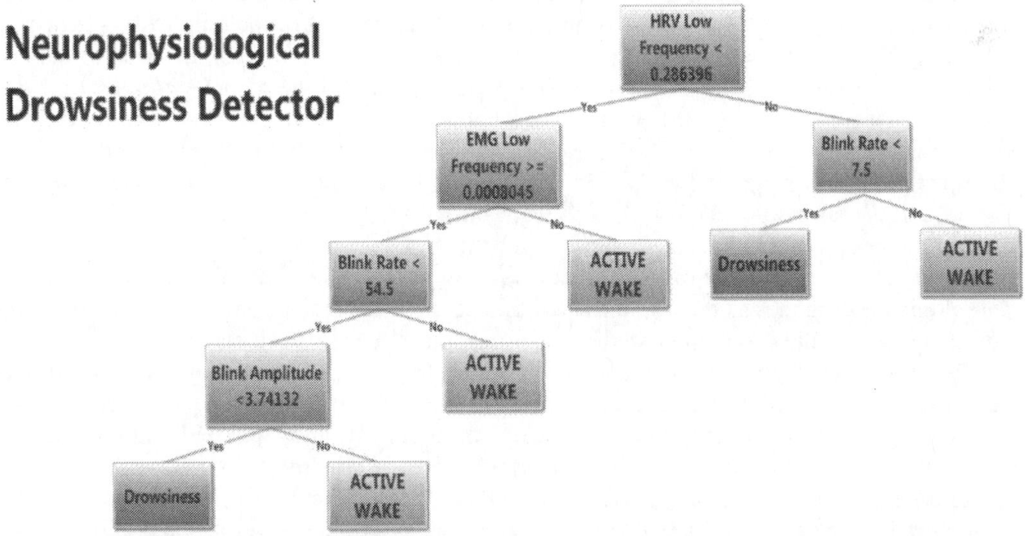

Figure. 3. Visualization of the rule-based algorithm (binary decision tree) for drowsiness detection based on the participants' heart-rate variability, chin electromyographic and eye blink activity.

5 Discussion & recommendations

The Airmanship 2.0. based framework is supported by all cases in the preliminary findings, and the typologies aided analysis and interpretation. In accordance with the literature on airline fatigue risk management, we explored the options that best fit the purpose of our methodology and the objectives of our project. These options satisfy the criteria of minimal intrusiveness and of operationalisability of tools for DSS development. Based on these criteria, Cogscreen AM was included in our deployed tools because of its low cost, ease of deployment and validity. In the Cogscreen sessions with Cases 1 and 3 inadequate scores indicate a degradation of cognitive abilities. Case 1 had over 30 years of professional experience in military and civil aviation, including the employment in different airlines, and the operation of various aircraft types with different approaches to the role of technology in the cockpit. Case 3 had 9 years of experience and mainly in the flight training industry before transiting to airlines performing better in the session but still producing low scores. Although Cogscreen AM was initially designed by the FAA based on military pilots, applications took place in samples of commercial pilots reporting a clear prediction of cockpit performance [18]. In cases where Cogscreen AM was applied in older groups of aviation employees the scores were lower as in Case 1. Our preliminary results, however, do not support the mediating effect of training and experience reported in other aviation professional groups such as air traffic controllers [19], showing a need for greater analysis on LoA. A re-evaluation of the assessment tool that takes into consideration the local population characteristics, including the attitudes of pilots and their airmanship, could be further explored to address such questions. In addition, as explained earlier (cf. Section 3.3), aeromedical examiners could create their own data banks. Similar concerns and propositions are reported where a multi-modal deep learning network can aid the classification of mental modes of pilots [20]. Nevertheless, in combination with the interview data, the Cogscreen results gain importance because these are in contrast with the perceived performance of Case 1 and Case 3. Both cases show confidence, yet in different systems; Case 1 in experience and Case 3 in automation. The latter supports the transition between generations from *aviate-navigate-communicate-manage* to *manage-communicate-navigate-aviate*.

Moreover, the sleep analysis of Cases 1 and 2 showed that there is a possibility of a sleep disorder. In Case 1, drowsiness lasted much longer than normal (almost up to 100 epochs), including 30 seconds outliers going from drowsiness to active wake. The lack of rest is evident (cf. Figures 1, 2), as N1 should have been within a range of 6-10 minutes and for Case 1 it was of 58.5 minutes, with lapses between N1 and drowsiness. Similarly, the participant went into N2 for less than the expected normal range of 1 to 2 hours. Importantly, the short periods of stages N3 and REM (i.e. N3 18 minutes and REM 19 minutes) indicate no rest and the fall into REM due to sleep deprivation. Similarly, Case 2 also shows signs of a sleep disorder. Although Case 2 slept longer than Case 1, short periods of REM are also observed, showing that the pilot did not rest. Because, however, we did not perform pathological examinations the possibility of this being an effect of apnea could not be ruled out. It is recommended that other studies examine apnea due to its effects. It is noteworthy to remind readers the Go! 2009 incident, when both pilots fell asleep during cruising altitude and the NTSB reported obstructive sleep apnea as the cause of the incident. Our preliminary report hence finds support in cases and concerns of pilot drowsiness [20]. In order, however, to be able to deploy the value of polysomnography from a research to an operational setting in a meaningful manner for fatigue risk management, a less time-consuming instrumentation is required. Decision trees are commonly used in aviation operations and have many similarities with other failure analysis trees applied in the industry, able to possibly gain greater acceptance by safety professionals. Furthermore, such data can aid peer pilot support programmes, extending the DSS for the enabling of awareness and self-declaration, and, if need be, referral.

However, although the classification of the proposed decision tree is scientifically acceptable (above 80%), its employment in aviation condition needs its further improvement. The latter could be achieved by including data from more participants (>=25), which will allow the algorithm's generalization. Then, more sophisticated machine learning algorithms (Support Vector Machines, neural networks) could be also employed to further improve the classification accuracy [21, 22]. In case of also employing brain activity data (e.g. wearable EEG devices), the number of estimated features would result in a big data perspective which may facilitate the use of deep learning algorithms [23]. The above steps may serve for a future roadmap towards the reliable integration of deep learning applications as onboard infrastructure for drowsiness estimation. The features used in the decision tree are in accordance with previous neuroscientific evidence. More specifically, the low frequency of the HRV timeseries is associated with the sympathetic branch of the autonomic nervous system. Drowsiness and sleep onset are closely related with the inhibition of sympathetic activity [24]. According to the AASM sleep scoring rules, drowsiness is mainly characterized by the onset of the alpha rhythm [25], which is associated by attenuated chin EMG and blink activity. Since, the features involved in the classification was only selected by the tree algorithm and not the developers, this is an evidence that the proposed methodology is of scientific and clinical validity. At operational levels, it would be easily implemented through microphones or cameras, whereas wearable devices with a bipolar electrode could provide more reliable, spectral EMG features.

6 Conclusion

Our paper marks a call for the selection of meaningful data and their sustainable use in air transport organisations. In the dynamic aviation environment, changes affect the pilots' willingness to report fatigue, their nature of work and their lifestyle. In this context, a pilot's capacity for information processing is radically decreased and their attention level diminishes, which may result in erroneous responses to crucial flight tasks. Following the Airmanship 2.0. dimensions, we reported the preliminary findings of a live project. The proposed systematic data collection method aims to promote evidence-based decision making and to address complex and least quantifiable pilot performance parameters, such as fatigue and sleep disorders. Our methodology employed both bio-signal processing and machine learning techniques to provide an objective estimator of drowsiness level. However, a greater amount of data is required in order to provide definitive results that could provide evidence-based feedback for policy making activities in the field of pilot fatigue. More advanced recording modalities such as EEG activity could be also employed to quantify the fatigue level prior entering a drowsiness state. However, we should bear in mind that a compromise among classification accuracy and unobtrusiveness should be made in order to result in applicable scientific solutions easily accepted by the aviation community. Further research should explore the relationship between boredom and drowsiness in the cruising altitude of flight with high LoA, the characteristics of pilots today and possible applications of such tools in peer pilot support programmes.

References

1. Besco, R.O. (1990). Subtle incapacitation of pilots: How to tell if your captain has died. Accident Prevention.

2. ECA (2012). Pilot fatigue. Available online at https://www.eurocockpit.be/sites/default/files/eca_barometer_on_pilot_fatigue_12_1107_f.pdf
3. Cropanzano, R. S., Ambrose, M. L., & Van Wagoner, P. (2019). 11 Organizational Justice and Workplace Emotion. Social Psychology and Justice.
4. Cahill, J., Cullen, P. and Gaynor, K. (2019). Pilot Wellbeing & Work-Related Stress (Wrs).27th International Symposium on Aviation Psychology.
5. Gawron, V. (2019). Automation in aviation – Guidelines. Mitre Technical Report MTR190018. McLean, VA: Mitre Corporation.
6. Levin, E., MR, M., COIMBRA, F., Keller, J. and Teo, A. (2019). Fatigue in Collegiate Aviation." International Journal of Aviation, Aeronautics, and Aerospace 6.4.
7. HELLERSTRÖM, D., Eriksson, E., ROMIG, E. and KLEMETS, T. (2010). Flight time limitations and fatigue risk management: comparison of three regulatory approaches", Boeing Flight Safety Foundation.
8. Hilditch, C.J., and Flynn-Evans, E.E. (2019). Controlled Rest: Profile of Use, Challenges, and Best Practices.
9. Rostaminia, S., Mayberry, A., Ganesan, D., Marlin, B. and Gummeson, J. (2017). iLid: Low-power Sensing of Fatigue and Drowsiness Measures on a Computational Eyeglass. Proceedings of the ACM on interactive, mobile, wearable and ubiquitous technologies 1(2). pp. 1-26.
10. O'Hagan, A.D., Issartel, J., Wall, A., Dunne, F., Boylan, P., Groeneweg, J., Herring, M., Campbell, M. and Warrington, G. (2019). Flying on empty: effects of sleep deprivation on pilot performance. Biological Rhythm Research, pp. 1-22.
11. Gateau, T., Ayaz, H. and Dehais, F. (2018). In silico vs. over the clouds: on-the-fly mental state estimation of aircraft pilots, using a functional near infrared spectroscopy-based passive-BCI. Frontiers in human neuroscience 12.
12. Cordeiro, R. and Carvalho J. (2019) Positive mental health in the workplace: a new global and preventive approach for flight staff. Авіаційна та екстремальна психологія у контексті технологічних. 3.
13. CAA, NLR, & NIN. (2018). Independent research on pilot fatigue measurement by the Netherlands Aerospace Center. Available at http://publicapps.caa.co.uk/docs/33/Pilot%20Fatigue%20Measurement%20Research_Final%20Report_v0.9.1clean.pdf.
14. Papanikou, M. (2017). The philosophy of safety: change and the re-organisation of the aviation industry. Ph.D. Thesis, University of Greenwich, UK.
15. Mohrmann, F., & Stoop, J. (2019). Airmanship 2.0: Innovating aviation human factors forensics to necessarily proactive role. International Society of Aviation Safety Investigators (ISASI). Annual Seminar.
16. Warr, P. (2001). Age and work behaviour: Physical attributes, cognitive abilities, knowledge, personality traits and motives. International review of industrial and organizational psychology, 16, 1-36.
17. Moser, D., Anderer, P., Gruber, G., Parapatics, S., Loretz, E., Boeck, M., ... & Saletu, B. (2009). Sleep classification according to AASM and Rechtschaffen & Kales: effects on sleep scoring parameters. Sleep, 32(2), 139-149.
18. Taylor, J., O'Hara R., Mumenthaler, M. and Yesavage, J. (2000). Relationship of CogScreen-AE to flight simulator performance and pilot age. Aviat Space Environ Med 7. pp. 373-80.
19. Taylor, J., O'Hara, R., Mumenthaler, M., Rosen, A. & Yesavage, J. (2005). Cognitive Ability, Expertise, and Age Differences in Following Air-Traffic Control Instructions. Psychology and Aging. 20(1). pp. 117-133.

20. Corbett, M.A. (2009). Science & Technology Watch: A Drowsiness Detection System for Pilots: Optalert®. Aviation, space, and environmental medicine, 80(2). pp.149-149.
21. Frantzidis, C. A., Bratsas, C., Papadelis, C. L., Konstantinidis, E., Pappas, C., & Bamidis, P. D. (2010). Toward emotion aware computing: an integrated approach using multichannel neurophysiological recordings and affective visual stimuli. IEEE Transactions on Information Technology in Biomedicine, 14(3), 589-597.
22. Spyrou, I. M., Frantzidis, C., Bratsas, C., Antoniou, I., & Bamidis, P. D. (2016). Geriatric depression symptoms coexisting with cognitive decline: a comparison of classification methodologies. *Biomedical Signal Processing and Control, 25*, 118-129.
23. Chriskos, P., Frantzidis, C. A., Gkivogkli, P. T., Bamidis, P. D., & Kourtidou-Papadeli, C. (2019). Automatic sleep staging employing convolutional neural networks and cortical connectivity images. IEEE transactions on neural networks and learning systems.
24. Reyes del Paso, G. A., Langewitz, W., Mulder, L. J., Van Roon, A., & Duschek, S. (2013). The utility of low frequency heart rate variability as an index of sympathetic cardiac tone: a review with emphasis on a reanalysis of previous studies. Psychophysiology, 50(5), 477-487.
25. Moser, D., Anderer, P., Gruber, G., Parapatics, S., Loretz, E., Boeck, M., ... & Saletu, B. (2009). Sleep classification according to AASM and Rechtschaffen & Kales: effects on sleep scoring parameters. Sleep, 32(2), 139-149.

MATEC Web of Conferences **314**, 01004 (2020) https://doi.org/10.1051/matecconf/202031401004
ICSC-ISATECH 2019

Relationship between air traffic demand, safety and complexity in high density airspace in Europe

Tamara Pejovic[1,*], *Fedja* Natjasov[2], and *Dusan* Crnogorac[3]
[1]EUROCONTROL, Performance Review Unit, Belgium
[2]University of Belgrade, Faculty of Transport and Traffic Engineering, Division of Airports and Air Traffic Safety, Serbia
[3]University of Belgrade, Faculty of Transport and Traffic Engineering, Computer Center, Serbia

Abstract. Air traffic performance of the European air traffic system depends not only on traffic demand but also on airspace structure and its traffic distribution. These structural (airspace structure) and flow characteristics (factors such as traffic volume, climbing/descending traffic, mix of aircraft type, military area activity) influence airspace complexity, which can affect controller workload and influence the probability of safety occurrence. In other words, all these dynamic and static complexity components can potentially have an impact upon the safety of the air traffic management (ATM) system. Having in mind fluctuation in traffic on daily, seasonal or annual level in certain airspace, a few questions arise: How changes in traffic demand influence complexity and conflict risk? Is there any correlation between traffic demand, conflict risk and complexity? Are there any differences between seasons? For that purpose, an investigation is performed on FAB Europe Central (FABEC) airspace, based on two weeks of operated traffic during the summer and winter of 2017. Air traffic complexity is estimated using the EUROCONTROL complexity metrics, while conflict risk is assessed using the conflict risk assessment simulation tool. Results show that certain positive relationship exists between traffic demand, conflict risk and complexity.

1 Introduction

In 2018, instrument flight rule (IFR) movements within the European airspace continued to grow strongly (4.65% versus 2017), making last year a new record year in terms of traffic volumes: the number of flights controlled reached an all-time record of more than 11 million [1]. The forecast growth indicates that by 2021, the European sky will handle over 12.3 million operations. This is an incredible challenge for the safety, the en route sector capacity and impact on the environment. The implementation of two operational concepts, the free route airspace (FRA) and functional airspace block (FAB), is seen as crucial 'tools' for solving those issues.

* Corresponding author: tamara.pejovic@eurocontrol.int

By definition, FRA is a specified airspace wherein users can freely plan a route between a defined entry point and a defined exit point, with the possibility of routing via intermediate (published or unpublished) waypoints, without reference to the air traffic service (ATS) route network, subject of course to availability. Within such airspace, flights remain subject to air traffic control (ATC) for the separation provision and flight level (FL) change authorizations.

The overall benefits of free route operations are distance and flight timesaving, resulting in less fuel consumption and a notable reduction of engine emissions, which benefits the environment [2]. FRA is seen as a cornerstone to improve FAB Europe Central (FABEC) structure and utilisation. From the other side, an implementation of FABs should bring further efficiency of airspace operations because FABs are 'based on operational requirements and established regardless of State boundaries, in which the provision of air navigation services and related ancillary functions are optimized and/or integrated' [3]. Currently, there are nine FABs established to cover almost the whole European airspace [3]. However, their implementation is still too slow (according to the European Commission [3]) causing inefficiency in the European ATM system.

1.1 Complexity of air traffic

Complexity of air traffic can be defined as the level of either perceived or actual spatial and time-related interactions between aircraft operating in a given airspace during a given period [4]. Specifically, complexity of air traffic in a given airspace can be very high solely because of the traffic intensity and its pattern in terms of mutual interactions between different traffic flows, as well as between individual aircrafts. Such presumably high complexity could be used for both planning and operational purposes mainly aimed at reducing it. Consequently, it may be reduced at the strategic, tactical and pre-tactical level. At each of these levels, it can have a spatial-based nature (such as airspace and airfield system design and/or assignment such as air routes, sectors, terminals, runway systems, etc.) and also time-based solutions (such as schedules, slot allocations, flow management, etc.). In that context, according to Netjasov et al. [4], complexity is understood as a demand characteristic of air traffic that is to be served by an appropriate supply system.

Traffic complexity affects control task complexity (Figure 1), where the control is performed by human operator [4]. It is expected that a more complex task will produce a higher workload. However, the workload influenced by task complexity differs between ATCos (Figure 1) due to differences in their working environment, perception of the traffic situation, personal experience, etc. Therefore, complexity represents a contributing factor of task complexity and ATCo workload.

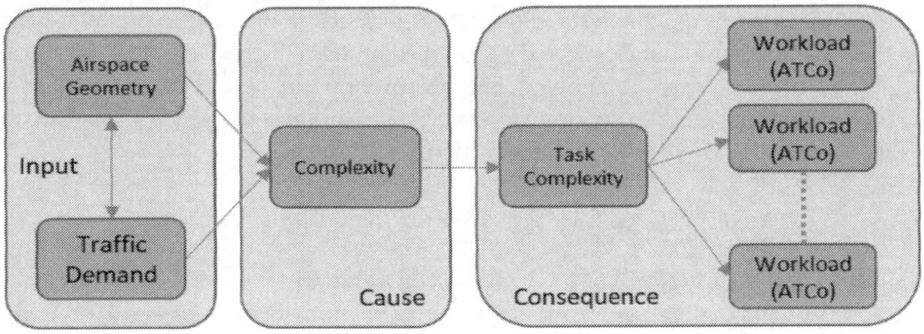

Figure. 1. Scheme of the relationship between complexity, task complexity and workload

The approach presented in this chapter is based on EUROCONTROL [5] methodology, with exclusion of ATCo workload issue from the explicit consideration. Approach is taking a macroscopic view, and it is considering four complexity components: adjusted density, potential vertical interactions, potential horizontal interactions and potential speed interactions. A single metric, 'complexity score', which incorporates these four separate parameters, was considered as the simplest for benchmarking purposes. Recently, Pejovic and Lazarovski [6] have studied the performance of the North European Free Route Airspace using EUROCONTROL approach.

1.2 Conflict risk

The International Civil Aviation Organisation (ICAO) has developed the Collision Risk Model (CRM) as a mathematical tool used in predicting the risk of mid-air collision [7]. Although aircraft collisions have actually been very rare events, contributing to a very small proportion of the total fatalities, they have always caused relatively strong impact mainly due to relatively large number of fatalities per single event and occasionally the complete destruction of the aircraft involved.

From other side, one of the principal matters of concern in the daily operation of civil aviation is the prevention of conflicts, i.e. Potential Losses of Separation (PLoS) between aircraft either while airborne or on the ground, which might escalate to collisions. A PLoS is a situation when two aircrafts come closer to each other than a specified minimum distance both in the horizontal and the vertical planes. One can imagine that PLoSs are more frequent event than collisions, so assessment of conflict risk is becoming important.

In order to determine whether or not PLoS situation exists and to calculate a conflict risk value, a cylinder-shaped 'forbidden volume' is defined around the aircraft [8]. A PLoS exists between two aircrafts if one of them enters the other's forbidden volume. They could be of a crossing or an overtaking type, depending on the aircraft trajectory relations both in horizontal and vertical planes [9].

Dealing with a conflict risk instead of a collision risk (a concept established by ICAO) is enabling a proactive safety approach, which is much closer to everyday ATCo activities.

2 Study approach

To analyse how future changes in airspace structure and traffic flow could influence complexity and safety performance, this paper proposes a showcase methodology on the analysis of FABEC. FABEC, which includes airspaces of six countries (Belgium, France, Germany, Luxembourg, the Netherlands and Switzerland) is one of the biggest Functional Airspace Blocks in Europe and is handling more than half of the European annual traffic. According to [10] this 'airspace is one of the busiest and most complex in the world' with 'most major European airports, major civil airways and military training areas located in this area'. Since June 2013, FABEC is officially in operation.

2.1 Traffic data and scenarios

Two traffic scenarios covered one week of summer (July 3-9, 2017, with 131.268 flights) and winter (November 13-17, 2017, with 94.947 flights). For each traffic scenario, calculation of complexity parameters (calculated using the NEST tool [11]) and safety indicators (calculated using the Conflict Risk Assessment Tool [9]) was done using the same input - summer and winter traffic.

2.2 Assessment of complexity and safety indicators

The assessment of complexity was done using the EUROCONTROL complexity score [5] as airspace complexity indicator. The two main metrics that define the complexity score are the "adjusted density" and the "structural index". The latter is derived from three parameters describing potential number of interactions in specific situations classified as vertical, horizontal and the mix of aircraft performances. The "adjusted density" assesses the potential interactions resulting from density, including uncertainty in the trajectories and time, while the "structural index" balances the density metrics according to the interaction geometry and aircraft performance differences. Formulas used for the calculation of complexity score are explained in [5].

The assessment of safety indicators was done using Conflict Risk Assessment Tool [9]. This tool is intended for the simulation of planned or analysis of realized air traffic, consisting of flight trajectories (4D trajectories) crossing a given airspace, with the aim of assessing safety performance. Conflict Risk Assessment Tool has been developed as a network based simulation model consisting of three modules, each being used for the calculation of certain safety indicators [9]: Separation violation detection module (dynamic 3D conflict detection model based on known flight intensions and distance-based separation minima - used for the calculation of duration, severity and number of PLoS), Traffic collision avoidance system activation module and Conflict risk assessment module. The safety minima separations used in analysis were: horizontal separation – 5 NM and vertical separation – 1000 ft; however those values were relaxed for 10% (4,5 NM and 900 ft) in order to deal with potential position and altitude inaccuracies.

3 Results

3.1 Complexity and safety indicators overall analysis

Daily fluctuations of both complexity and safety indicators follow similar pattern throughout the week in both summer and winter. Traffic values indicate that traffic demand during winter is lower than during summer in a range from 22 to 37%. Similarly the complexity score values fluctuate - summer values are higher in a range between 17 and 29%. Overall, changes of daily complexity values are following the changes of daily traffic demand.

Contrary, the changes in daily number of PLoS and conflict risk do not follow strictly the changes in daily traffic demand. However, variations in the number of PLoS are following the changes in conflict risk. The number of PLoS during winter is lower in a range from 33 to 63%. Similarly, the conflict risk shows lower values in winter in a range from 28 to 65%.

Complexity analysis shows that total hours of interactions (bars, Figure 2 up) increase during winter mainly due to increase in speed interactions (yellow bars, Figure 2 up) which could indicate the greater changes in the mix of aircraft used. However, overall complexity score reduces during winter by 15–20% depending on the day of the week. This indicates that overall complexity score (black line, Figure 2 up) is mainly influenced by changes in adjusted density (green line, Figure 2 up). Adjusted density assesses aircraft interactions resulting from density, including uncertainty in the trajectories and time. Adjusted density shows that interactions are not only related with the traffic volume, however also with how this traffic is dispersed in airspace.

The analysis of number of interactions and number of PLoS per hour of flight (Figure 2 down) shows that the total number of interactions per hour of flight reduces during winter by over 23% (0.224 in summer vs. 0.172 in winter). The number of PLoS per hour of flight is somewhat more stable and does not change much with decrease in traffic. Overall change is approximately 13% (0.015 in summer vs. 0.013 in winter).

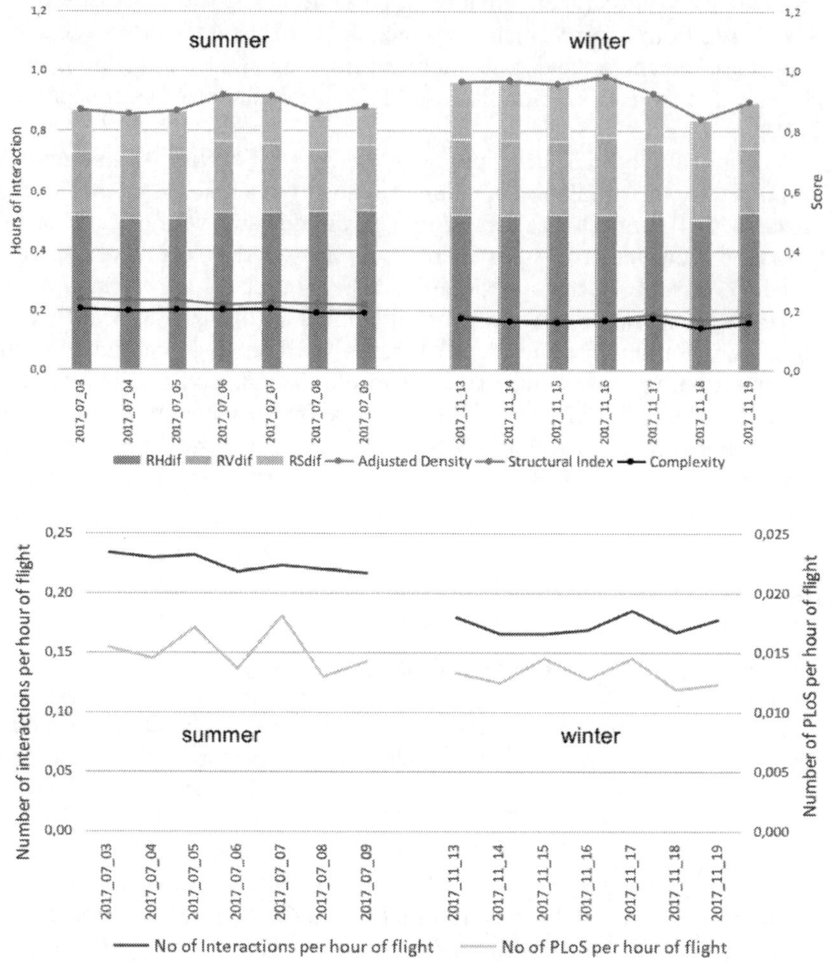

Figure. 2. Complexity parameters (up) and number of aircraft interactions and PLoS per hour of flight (down)

3.2 Correlation between traffic, complexity and safety indicators

A very strong correlation ($R^2 = 0.9807$) was found between the daily number of flights and complexity (Table 1). Strong correlations were also found between safety indicators (slightly better correlation with the number of PLoS) and the total number of flights. Similarly, strong correlations exist between safety indicators and complexity.

These findings could lead to a conclusion that with increase in traffic, one can expect the higher complexity, followed by the higher number of PLoS and conflict risk. In other words, this means that ATCo task load will increase, leading to a higher ATCo workload.

Table 1. Linear correlation coefficients for both seasons combined

Both seasons	Complexity	PLoS	Conflict Risk
# Flights	+0.9807	+0.8819	+0.8008
Complexity	-	+0.9138	+0.8296

Seasonal comparison - the results of correlation analysis between traffic demand, complexity and safety indicators (conflict risk) show the positive correlation between the number of flights and complexity score (as dependent variable) that is stronger in winter (summer $R^2 = 0.7163$ (Table 2) vs. winter $R^2 = 0.9022$ (Table 3)). This is expected as daily complexity scores follow the daily number of flights. Moreover this could be explained by the fact that during the winter traffic is more uniform, while during summer there are more charter and business flights (unscheduled flights) that could change traffic flows and locations of PLos points, which in turn is decreasing predictability and increasing complexity score. Operationally, this could also potentially increase ATCo's workload during summer.

Table 2. Linear correlation coefficients for summer season

Summer	Complexity	PLoS	Conflict Risk
# Flights	+0.7163	+0.3716	+0.3114
Complexity	-	+0.6980	+0.5144

Table 3. Linear correlation coefficients for winter season

Winter	Complexity	PLoS	Conflict Risk
# Flights	+0.9022	+0.5005	+0.2207
Complexity	-	+0.5640	+0.2843

Moreover, the positive correlation between the number of flights and conflict risk (as dependent variable) is not strong (in both seasons $R^2<0.5$, although in summer is somewhat stronger). Similarly, the positive correlation between complexity score and conflict risk is not strong (R^2 is higher in summer than in winter).

Correlation between the number of flights and number of PLoSs is not strong (although somewhat higher in winter). Contrary, correlation between complexity scores and the number of PLoS shows a positive correlation (stronger in summer).

Overall, it can be concluded that correlation between complexity and the number of PLoSs is stronger than between complexity and conflict risk (Tables 2 and 3). Similar behaviour can be observed in the case of correlation between the number of flights and the number of PLoS. However, it has to be noted that the conflict risk as an indicator contains more information about the loss of separation than just the total number.

3.3 Analysis of potential losses of separation

3.3.1 PLoS duration and severity

Each PLoS is characterised by the combination of severity (related to the breach of separation) and duration. The severity of the PLoS depends on the minimum distance (spacing) between the pair of aircraft (S_{min}) and the applied separation minima (Sep_{min})). The severity presents the level of aircraft proximity and is defined either for the violation of separation in the horizontal or the vertical plane, or both [8]: Severity = (Sep_{min} - S_{min})/ Sep_{min}; where 0 < = Severity < = 1.

Severity could be 1 in the case when both aircraft are at the same point in the horizontal plane (although they could be properly separated vertically) or in the case when both aircraft are at the same altitude (however they could be properly separated horizontally).

Results of PLoS duration analysis show that majority of PLoSs are short, up to 30 sec (roughly two-thirds, i.e. 372 cases), while almost 90% do not last more than 1 min. Results of severity analysis show that in 80% of cases severity is 1, which means that both aircraft were at either the same flight level but not in the same point in the horizontal plane or at the same point in the horizontal plane but at different flight levels.

3.3.2 Horizontal and vertical separation at closest point of approach

The results of the distribution of minimum vertical separation at the closest point of approach (CPA) show that almost 80% of PLoSs are at the same flight level or are separated vertically up to 100 ft. The results of horizontal distribution show that roughly 50% of PLoSs have breach of less than 3 NM.

3.4 Complexity and safety per flight level

Both complexity and conflict risk can change with flight level. The distribution of an average daily complexity and average daily number of PLoS per FLs is shown on Figure 3.

The highest average values of complexity are on higher altitudes (FL350 to FL380) which correspond to the level used for en route cruise. Somewhat increased values of complexity could be also seen between FL220 and FL240 (corresponds to situations in which flights are entering or leaving lower airspace), however, the number of PLoSs is not increased at this level band.

Distributions of average daily complexity and average daily number of PLoS per FL are similar, with lower values during winter days (Figure 3). Additionally, it can be concluded that traffic demand is influencing higher complexity values; moreover, the number of PLoSs evidently correlates with higher complexity values (Figure 3).

Figure 3 shows that in the summer period increase in the number of PLoS at high complexity altitudes is of higher magnitude than in winter months. This could be related to the fact that summer traffic is less predictable (due to the existence of increased number of charter flights and summer destinations traffic).

4 Conclusion

Air traffic performance of the European air traffic system depends on traffic demand but also on airspace structure and its traffic distribution. These structural and flow characteristics influence airspace complexity, which can affect controller workload and influence the probability of safety occurrence.

An investigation is performed on FABEC airspace in Europe, based on 2 weeks of realised traffic during summer and winter of 2017, with aim to answer several questions: How changes in traffic demand influence complexity and conflict risk? Is there any correlation between traffic demand, conflict risk and complexity? Are there any differences between seasons?

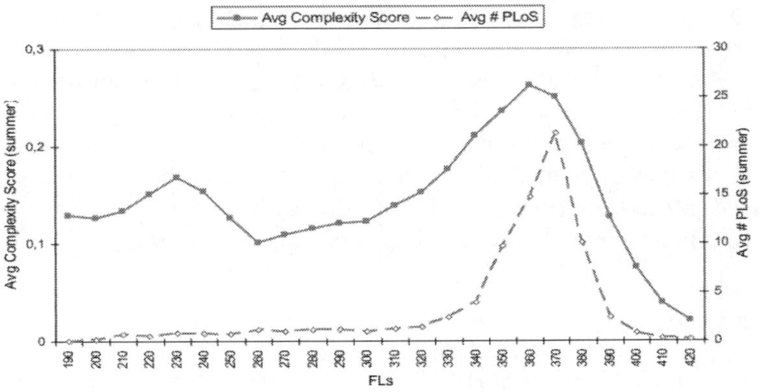

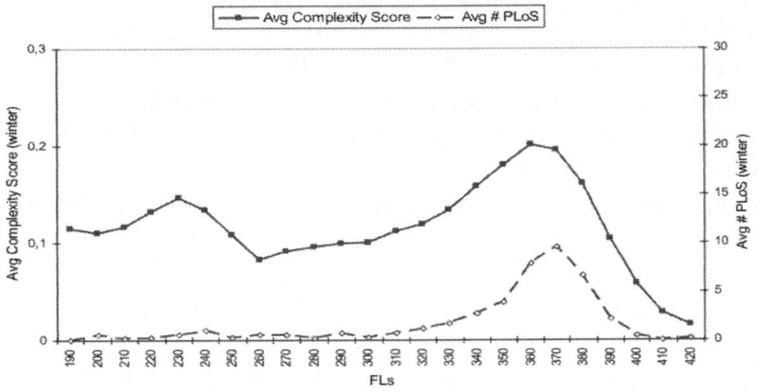

Figure. 3. Distribution of complexity and the number of PLoS per FL (red line – summer/blue line – winter)

Daily fluctuations of both complexity and safety indicators follow a similar pattern throughout the week in both summer and winter. Analysis of complexity parameters shows that overall complexity score is mainly influenced by changes in adjusted density which show that interactions are not only related with the traffic volume but also with how this traffic is dispersed in airspace.

The changes in the number of PLoS and conflict risk do not follow strictly the changes in daily traffic demand, and the numbers of PLoS and the conflict risk are lower in winter. This could be related to the fact that traffic demand is lower in winter months and that traffic is more predictable.

Strong correlations were found between traffic demand, safety and complexity indicators. These findings could lead to conclusion that with increase in traffic, one can expect the higher complexity, which in turn influences the number of PLoS and conflict risk. In other words, this means that ATCo task load will increase, leading to a higher ATCo workload.

Both complexity and conflict risk can change with flight level. The highest average values of complexity and number of PLoS are on higher altitudes (FL350 to FL380) which correspond to the level used for en route cruising. Increase in number of PLoS at these altitudes is higher, in relation to increase in complexity, during summer. This could be related to the fact that summer traffic is less predictable (due to existence of increased number of charter flights and summer destinations traffic).

In a conclusion, this small-scale analysis showed that changes in traffic demand do influence complexity and safety performance (both in terms of the number of PLoS and conflict risk). Moreover, this analysis set a benchmark for future monitoring of safety and operational performance after FRA implementation at FABEC airspace. Further analysis should investigate whether dispersion of traffic after FRA implementation is enough to create complexity decrease and whether change in complexity have not compromised safety and ATCo workload. Moreover, analysis could increase credibility by considering traffic flows, sectors, types of flights (charter, low cost, business, etc.) and vertical profiles of flights.

References

1. Seven-year Forecast February 2019. EUROCONTROL, Brussels, Belgium, 2019.
2. S. Aneeka, Z. Zhong, "NOX and CO2 emissions from current air traffic in ASEAN region and benefits of free route airspace implementation". Journal of Applied and Physical Sciences. 2016; 2(2):32-36. DOI: 10.20474/japs-2.2.1
3. Functional airspace blocks. European Commission, Brussels, Belgium, 2019. Available: https://ec.europa.eu/transport/modes/air/single-european-sky/functional-airspace-blocks-fabs_en [Accessed 24 July, 2019].
4. F. Netjasov, M. Janic, V. Tosic, "Developing a generic metric of terminal airspace traffic complexity". Transportmetrica. 2011; 7(5):369-394. DOI: 10.1080/18128602.2010.505590
5. Complexity Metrics for ANSP Benchmarking Analysis. EUROCONTROL, Brussels, Belgium, 2006.
6. T. Pejovic, A. Lazarovski, "Cross border Free Route Airspace implementation – performance overview methodology". In: Proceedings of 23rd World Conference of the Air Transport Research Society (ATRS); 2019; Amsterdam, The Netherlands
7. CIRCULAR 319 - A Unified Framework for Collision Risk Modelling in Support of the Manual on Airspace Planning Methodology for the Determination of Separation Minima (Doc 9689), International Civil Aviation Organization, Montreal, Canada, 2009.
8. F. Netjasov, "Framework for airspace planning and design based on conflict risk assessment, Part 2: Conflict risk assessment model for airspace tactical planning". Transportation Research Part C. 2012; 24:213–226. DOI: 10.1016/j.trc.2012.03.003
9. F. Netjasov, D. Crnogorac, G. Pavlovic, "Potential safety occurrences as indicators of air traffic management safety performance: A network based simulation model". Transportation Research Part C. 2019; 102:490–508. DOI: 10.1016/j.trc.2019.03.026
10. Functional Airspace Block Europe Central (FABEC). Available: https://www.fabec.eu/ [Accessed 24 July, 2019]
11. NEST (Network Strategy Tool) User Manual. EUROCONTROL, Brussels, Belgium, 2017.

Fedja Netjasov's work was conducted with support from the Project number 36033 commissioned by the Ministry of Education, Science and Technological development of the Republic of Serbia.

Expediency of ATO in heavy rail: A survey for the Dutch Railways

M. Akbari[1,*], *B.* Hoogewoonink[2], *B.* Godziejewsk[3], and *M.* Rajabalinejad[4]
[1]Master student, (University of Twente, ME, the Netherlands)
[2]Consultant Railway Signalling and Safety, (Mott MacDonald, the Netherlands)
[3]Senior Rail Manager, (Mott MacDonald, the Netherlands)
[4]Assistant Professor, (University of Twente, ME, the Netherlands)

Abstract. Automatic train operation (ATO) is a relatively new trend in heavy rail. While technical specifications define system and function requirements, implementation and adoption of ATO requires a business case. In this research we have developed a framework which could indicate if a corridor is suitable for automatic operation or not. We have approached this challenge by using product development methods such as QFD (Quality Function Deployment) and methods developed specifically for this study. The results indicate that ATO could influence performance, time to market and costs with respect to manual operation. The influence depends on the characteristics of a corridor, corridor segmenting as well as the operational procedures. The positive influence of ATO, considering the current state of the Dutch railways, could be limited and there may be additional challenges.

1 Abbreviations

ABBREVIATION	FULL FORM
AoE	ATO over ETCS
ARCC	Automated Rail Cargo Consortium
ATO	Automatic Train Operation
DMI	Driver Machine Interface
EEIG	European Economic Interest Grouping
ERA	European Union Agency for Railways
ERTMS	European Railway Traffic Management System
ETCS	European Train Control System
GoA	Grade of Automation
IEC	International Electrotechnical Commission
JU	Joint Undertaking
KPI	Key Performance Indicators
QFD	Quality Function Deployment
SPAD	Signal Passed At Danger

* Corresponding author: maryam.akbari@alstomgroup.com

2 Introduction

Autonomous and automatic operation are the new trends in land transportation. Around 25 cities in the world have driverless metro systems and self-driving cars are expected soon. This trend did not find its way to heavy rail until recently. In the late 2017, the first fully autonomous freight train started operation in a mining railway line in Australia.

This project along with the deployment of ERTMS have perhaps encouraged heavy rail undertakings to consider ATO. Within the developments of Shift2rail, now more demonstrators are being planned and there are a few ATO related projects (e.g. X2Rail, ARCC) [1]. In addition, ERA's interface and system specifications for an interoperable ATO over ETCS (GoA 2) operation are expected to be released soon.

However, implementing ATO on a rail corridor requires a business case. Corridors with special characteristics could be more suitable for ATO as benefits become significant while there could be corridors which benefits of ATO are limited. Consequently, it appears that a framework which can aid in the decision-making process is necessary. The objective of this research is to develop a tool which can be used in early phases of decision making. The tool is expected to determine the expediency of automatic operation for rail corridors in the Netherlands.

In this paper the methodology to develop and validate a decision-making tool is defined. This tool can be used to determine a business case for implementing ATO. The case studies were done by applying the tool to several rail corridors in the Netherlands. The conclusion and recommendations for future work were drawn based on study results.

3 ATO over ETCS

AoE is a concept in which ATO drives the train while a safety critical system (ETCS) is in place and protecting the train from hazards. ETCS is viewed a solution to implement an affordable and interoperable ATO in the heavy rail. The projects done by EEIG and Shift2Rail JU have contributed to better understanding the challenges related to ATO and have support elaboration of requirements specifications. The existing specifications for AoE are applicable for ETCS levels 1, 2 and 3.

3.1 ETCS

ETCS is an interoperable train control system promoted by the European Commission. ETCS Level 2, which will be applied in the Netherlands, is based on cab signalling and continuous track to train data transmission. It provides an inherently safe operational environment for the movement of trains throughout the railway network [2].

3.2 ATO

ATO is a non-safety critical system which performs all or part of the tasks assigned to the train driver. The tasks performed by ATO depend on the GoA. IEC has defined 4 grades of automation [3]. This is further detailed in Table 1.

Table 1. Share of tasks between the train driver and the System

Basic functions of train operation	**GoA 1**	**GoA 2**	**GoA 3**	**GoA 4**
Ensure safe movement of train	ETCS	ETCS	ETCS	ETCS
Drive train	D	S	S	S
Supervise passenger transfer	D/ A	D/ A	A	S
Operate a train	D	D	A	S
Ensure detection and management of emergency situations	D	D	A	S

* D: Driver, A: Assistant, S: ATO

3.3 Current AoE development

Only a few ATO over ETCS projects in heavy rail have been executed around the world. These projects, as shown in Table 2.

Table 2. ATO over ETCS Deployment in the World

Parameter	China Intercity	Thameslink	Toluca – Mexico City	Iron ore train
GoA level	2	2	2	4
ATP	Eq. to ETCS L1	ETCS L 2	ETCS L2	ETCS L2

3.4 Other ATO developments

Although ATO over ETCS is most promising development, however various railway undertakings and infrastructure managers have and are considering ATO over their legacy ATP systems. The slow deployment of ETCS has motivated these organizations to look for solutions to integrate automation within their legacy ATO systems. Example of such developments are:

1. France, remote cab for freight corridors,
2. Germany, remote cab for freight corridors,
3. The Netherlands, GoA 2 on passenger corridors,
4. Austria, GoA 4 for freight corridors,
5. Russia, GoA 2 for passenger corridor,
6. Russia, remote cab for freight corridor,

4 Methodology

In this section, the main assumptions and approach for developing the framework are briefly discussed.

4.1 Assumptions

This research focuses on ATO over ETCS Level 2. As the target train control system for Dutch Railways is expected to be ETCS Level 2 [4].

It is assumed that in GoA 3 and 4 there are obstacle detectors installed on-board. Furthermore, the performance of these devices is assumed to be at least similar to a human driver by the time the ERA specifications for GoA 3 and 4 are released.

4.2 Approach

The approach taken to develop the framework is summarized in Figure 1 and explained below.

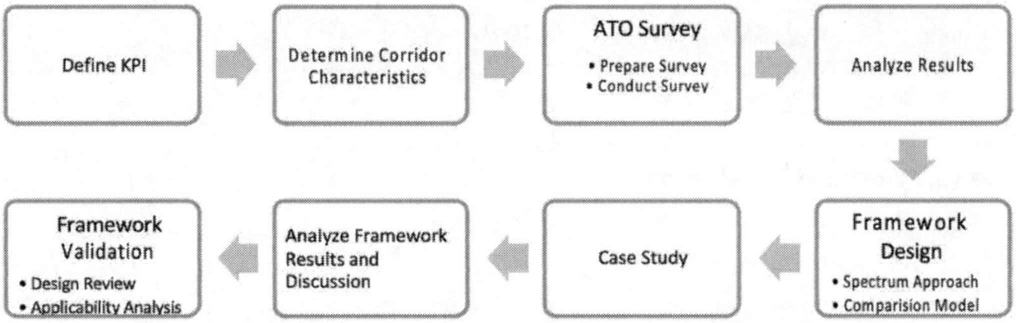

Figure. 1. Summary of Methodology

The initial step to develop the decision-making framework was to define KPIs which can support the business case. Considering the model developed by Hayes and Wheelwright [5], several KPIs along with their contributing indicators which best represented performance, time to market and financial aspects were defined.

Then 16 corridor characteristics which could influence the defined KPIs, were identified. The relation between the KPIs and corridor characteristics as well as the impact of different GoA on the KPIs were determined by Dutch railway experts using a survey. In the next step, the survey was prepared, conducted and results were analysed. The survey results were used to form a multicriteria analysis to design the decision-making framework.

The ATO decision-making framework was then applied to several rail corridors in the Netherlands and results were analyzed and discussed.

Finally, in order to validate the framework two methods were developed. The first validation method included review meetings. As the final step, the framework was applied to two currently automated urban rail systems.

5 Defining KPIs

KPIs are set of quantifiable measures that organizations use to measure their performance against business objectives. As "performance" is viewed differently within every organization, different measures are used by them.

However, the KPIs defined for this research are typical in the railway environment and often used by railway undertakings and infrastructure managers.

5.1 Performance

The performance parameters generally include those parameters which are part of the concessions and deviation from agreed values results in penalties. These KPIs are:

5.1.1 Safety

Safety is defined as the freedom from conditions that can cause death, injury, occupational illness, damage to or loss of equipment or property, or damage to the environment [6]. The

European parliament and of the council have defined several indicators to measure safety in the railways in Directive (EU) 2016/798 [7]. These indicators include:

- No. of SPADs,
- No. of over-speeding,
- No. of collisions with object / person near the track,
- No. of collisions with object / person on the platform,
- No. of collisions with road vehicles at level crossings,

5.1.2 Punctuality

Punctuality is the property of an event occurring "on time". The main contributing indicator for punctuality in heavy rail is variation from running time.

5.1.3 Capacity of a corridor

"Capacity is a measure of the ability to move a specific amount of traffic over a defined rail line with a given set of resources under a specific service plan" [8]. Therefore, capacity of a corridor can be measured through the difference between theoretical and practical capacity. The former refers to the maximum number of trains which can run over a route in a strictly perfect, mathematically generated environment. The latter represents the practical number of trains which can be moved considering a reasonable level of reliability [8].

5.1.4 QoS (Quality of Service)

Ride comfort is a common measure to determine the quality of service. For this study, vibration is selected as a representative parameter. Braking and acceleration contribute to the felt vibration and could be influenced by automation. Therefore, the contributing indicator for QoS is decided as the percentage of inconvenient braking / accelerating.

5.1.5 Reputation

Reputation, by definition of the Cambridge dictionary, is the opinion that people in general have about someone or something. Innovation is one of the contributing factors to reputation, [9]. For this research innovation is measured by percentage of automated driving tasks.

5.2 Time to market

TTM refers to the time from which a company initially conceives a product or service idea to the point when the actual product or service is accessible in the market. In the railways, the time to market could includes flexibility as it deals with the train service becoming available to the passengers.

5.2.1 Flexibility

Flexibility could be referred to making changes. A flexible system is a system in which changes are relatively easy to make [10].

Scheduling a train driver is one of the most time taking and complex tasks for the railway undertaking. There is little flexibility in this process and changes are difficult to make due to the numerous variables and constraints. Example of the variables and constraints include qualifications, legislations, shift hours, etc. [11]. In order to measure flexibility, the number of variables to plan a driver for train operation is used.

5.3 Financial aspects

Financial and economic aspects are related to the life cycle costs which includes the investment, operation and maintenance costs.

5.3.1 Railway operational expenses

For this study, the operational expenses are related to the operation of the train in the railway. In the railways, the operation charges are mainly due to the energy consumption and the costs associated with the driver / conductor. Accordingly, the following could represent railway operating expenses:

5.1 Energy usage per train per km
5.2 Staff expenses per train

5.3.2 ATO Investment expenses

Capital costs are fixed, one-time expenses for purchasing equipment. For this study, the capital costs are expressed as the costs to deploy an on-board and trackside ATO equipment.

5.3.3 Railway maintenance expenses

Railway infrastructure and rolling stock require frequent maintenance in order to keep them in good condition, ensure smooth running of the trains and avoid hazardous situations. The maintenance costs could be divided between and measured with the expenses related to the infrastructure manager and the railway undertaking

6 Determining corridor characteristics

This section includes the corridor characteristics assumed to have an influence on the KPIs discussed in the previous section. These corridor characteristics were determined during brainstorming sessions with Dutch railway experts as well studying available literature, as shown in Table 3.

Table 3. Form Determined Corridor Characteristics

No.	Corridor Characteristic	No.	Corridor Characteristic
1	Average Distance between stations	9	No. of Level Crossings (LX) for Road Vehicles
2	Number of Stations	10	Available Infrastructure Capacity
3	Maximum Line speed	11	Infrastructure Condition

No.	Corridor Characteristic	No.	Corridor Characteristic
4	No. Junctions	12	Traffic Heterogeneity
5	Suicide history	13	Current No. of trains
6	Corridor Length	14	Train Type
7	Track Segregation	15	Number of Passengers
8	Average Track Adhesion Conditions	16	Signalling System

7 ATO survey

In this section, the process to prepare, conduct and analyze the results of the survey are explained.

7.1 Preparation of the ATO survey

Selecting a template format and a ranking system were required to perform the survey. The survey template was taken from the relation matrix of the QFD, which is a structured approach to define customer needs.

Two qualitative ranking steps were defined. The choice of a qualitative over quantitative approach was done to have less inconsistent data and to assure that the survey could be completed in a realistic timeframe. The first ranking step, used to indicate the relation between the KPIs and the corridor characteristics, was a binary or 0 – 1. Experts could state if there is a link between corridor characteristics and a KPI or not. The second one would indicate how KPIs in different GoA would vary in relation to GoA 1. Table 4 presents the choices given to the experts for the second step:

Table 4. Second ranking step

Ranking	Symbol
Strong increase (> 50%)	↗↗
Weak increase (< 50%)	↗
No Change	→
Unknown	?
Weak decrease (< 50%)	↙
Strong decrease (> 50%)	↙↙

The survey which given to the participants can be found in Appendix A.

7.2 Conducting the ATO survey

15 candidates for conducting the survey were selected from different organisations to get representative and comprehensive results. Experts from infrastructure managers, railway

undertakings, signaling consultants, university scholars and governments representatives were asked to participate. Experts were also asked to indicate the KPIs which are applicable in a future business case.

The list of experts whom were asked to participate, their expertise and related organizations can be found in Appendix B.

7.3 Integrating survey results

An integration criterion was needed to combine and integrate the different viewpoints experts could have. The following criterion was used to merge the answers:

"A marking in the relation matrix is decided for if at least half of the participants have marked it."

8 ATO survey results

After the survey answers were collected and integrated, they were reviewed. The answers provided by the experts were similar in most cases. Dissimilar answers were given for the impact of automation on safety and maintenance KPIs. In this section, the results of the survey are detailed and discussed.

8.1 Business case KPIs

As mentioned in 7.2, experts were asked to specify which KPIs should be taken into consideration for a business case. Among the initially defined KPIs, experts selected 7 KPIs, as shown in Table 5. Some KPIs were linked to corridor characteristics, while others were independent from the characteristics of a corridor. This is shown in the last column of Table 5.

Table 5. Business Case KPIs

Category	KPI& Contributing Indicators	GoA 2	GoA 3	GoA 4	Dependency
Performance	1. Punctuality				
	1.1 Variation from running time	↘	↘	↘	Corridor dependant
	2. Capacity of a Corridor				
	2.1 Difference between theoretical and practical capacity	↘	↘	↘	Corridor dependant
	3. Reputation				
	3.1 Percentage of automated driving tasks	↗	↗↗	↗↗	Corridor independent
Time to Market	4 Flexibility				
	4.1 Number of variables to plan a driver for train operation	→	↘	↘↘	Corridor Independent
	5 Railway Operational Expenses				
	5.1 Energy usage per train per km	↘	↘	↘	Corridor dependant

Category	KPI& Contributing Indicators	GoA			Dependency
		2	3	4	
Financial Aspects	5.2 Staff expenses per train	→	↘	↘ ↘	Corridor independent
	6 ATO Investment Expenses				
	6.1 On-board ATO	↗	↗ ↗	↗ ↗	Corridor dependant
	6.2 Trackside ATO	↗	↗↗	↗↗	Corridor dependant
	7 Railway Maintenance Expenses				
	7.1 Expenses for Infrastructure Manager	↗	↗	↗	Corridor dependant
	7.2 Expenses for Railway Undertaking	↗	↗	↗	Corridor dependant

8.2 Influencing corridor characteristics

Among the 16 corridor characteristics, only 8 have an influence on KPIs indicated in Table 5. These corridor characteristics are:

1. Number of stations
2. Track segregation
3. Theoretical number of trains (per hour)
4. Current number of trains (per hour)
5. Corridor length
6. Type of service (regional, intercity, freight, mixed)
7. Maintenance frequency of infrastructure
8. Total number of rollingstock

9 Framework design

The survey results indicate that there is a link between the corridor characteristics and the KPIs. ATO could impact the KPIs due to these links. Therefore, the state of the KPIs in GoA 1 (for a corridor) should first be determined. Then it could be decided if automation can have an impact on the KPIs or not. This is done by a method specifically developed for this research, the "Spectrum approach", as explained below.

9.1 Spectrum approach

The spectrum approach uses a binary logic to decide if a contributing indicator of a KPI is likely to impact the KPI or not. Every contributing indicator has its own spectrum. The spectrum depends on the corridor characteristics which have a link to the contributing indicator of a KPI. This is further explained with an example.

"Variation from dwell time" is a contributing indicator for punctuality. A variation happens when passenger boarding, and de-boarding does not take place according to the timetable. Conductors usually wait for all the passengers to de-board and board from a train, then lock the doors and allow the train to depart. The delays on the passenger rail corridors in the Netherlands were reviewed and variations were found to be common on busy stations. Examples of such stations include Amsterdam Central and Utrecht Central.

In GoA 4, the conductor is replaced with a system. The system no longer waits for the passengers, and starts the train as planned in the timetable. Therefore, the variation from dwell time is expected to decrease. However, this impact would only be noticeable at busy stations, where the variation from dwell time was originally more probable.

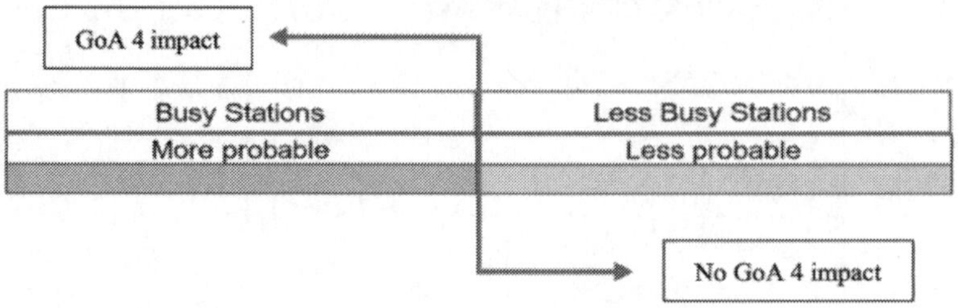

Figure. 2. Example of Spectrum Approach

9.2 Comparison model

In order to decide on the expediency of ATO, a quantitative cost – benefit analysis was done. The cost – benefit parameters were taken from the KPIs shown in Table 3. 4 possible states could result from the cost – benefit analysis:

- High costs – low benefits
- High costs – high benefits
- Low costs – low benefits
- Low costs – high benefits

These states could be translated to “Suitable for ATO”, “ATO not recommended” and “Unknown / Undecided”. For every GoA, a separate basis for comparison was defined.

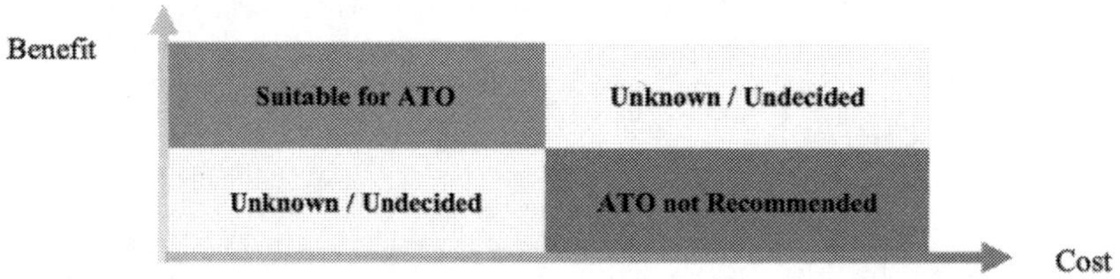

Figure. 3. Cost - Benefit Comparison Chart

10 Case study

8 different rail corridors in the Netherlands were selected as case studies for the application of the decision-making framework. The selected corridors are shown in table 6.

Table 6. Case studies.

No.	From	To	Service
1	Enschede	Den Haag	Mixed
2	Utrecht Central	Uitgeest	Mixed
3	Groningen	Zuidhorn	Regional
4	Betuweroute	-	Freight
5	Amsterdam Central	Utrecht Central	Intercity
6	Rotterdam -	Den Haag	Mixed

No.	From	To	Service
7	Arnhem -	Winterswijk	Regional
8	Nijmegen	Den Bosch	Mixed

The above rail corridors were evaluated with the support of the decision-making framework for their suitability to deploy ATO.

11 Framework results

The main objective of this research was to develop a tool which could suggest expediency of ATO over ETCS in the heavy rail. The methodology and approach taken during this research resulted in a framework which fulfilled the objective. In this section, the results of the case studies are presented and discussed.

11.1 Expediency of ATO for Dutch Railways

The case study results are shown in below figures. The vertical axis represents the benefits while the horizontal one is for the costs. The legend of Figure 3 applies to Figures 4, 5 and 6. A colour coding was used to distinguish between the different services, as detailed in Table 7.

Table 7. Colour Coding for Service

Service	Colour
Intercity	
Regional	
Freight	
Mixed	

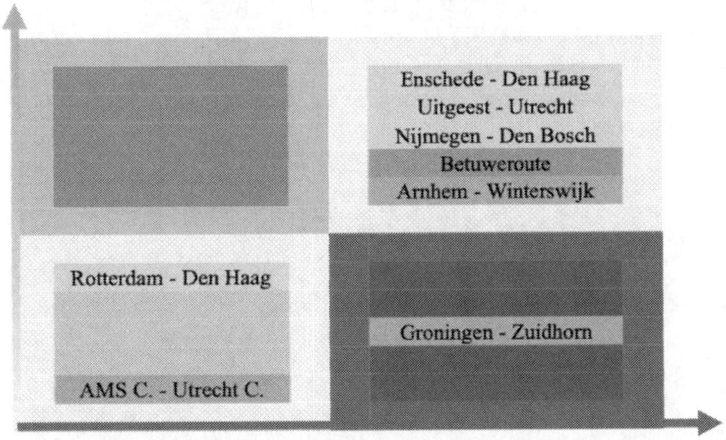

Figure. 4. Framework results for GoA 2

In GoA 2, there is no significant gain from the corridor independent KPIs (flexibility, reputation and staff costs). The corridor dependent KPIs (punctuality and energy usage) become noticeable on corridors where the costs are also high. Therefore, there is doubt about the suitability of GoA 2 for the heavy rail.

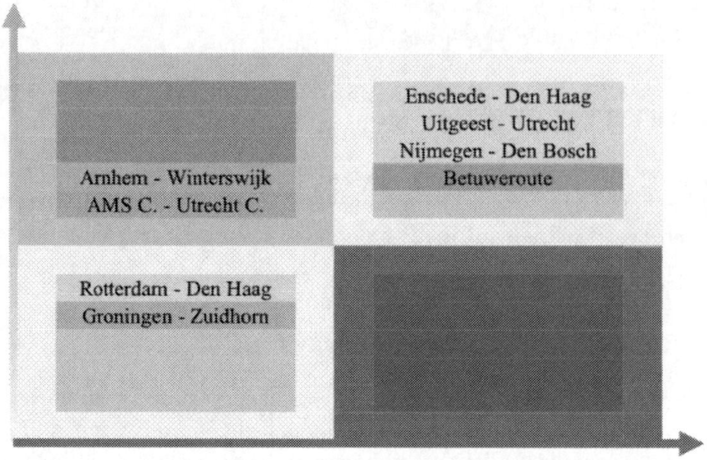

Figure. 5. Framework results for GoA 3

On corridors which are segregated, the investment cost for the on-board are not considerable and would be the same as GoA 2 as there is no requirement for obstacle detectors. For relatively short rail corridors, the investment cost for ATO trackside is also not remarkable. The investment costs for ATO are also not significant on regional corridors with light traffic. In case of GoA 3, the staff costs are partly reduced, and flexibility and reputation are gained to some extent. Two categories of corridors could be suitable for GoA main segregated corridors which are relatively short and regional corridors which include many stations. On these regional corridors, punctuality can be improved, and energy could be more efficiently used.

On the other rail corridors, there is doubt if the benefits from implementing ATO could compensate for the costs.

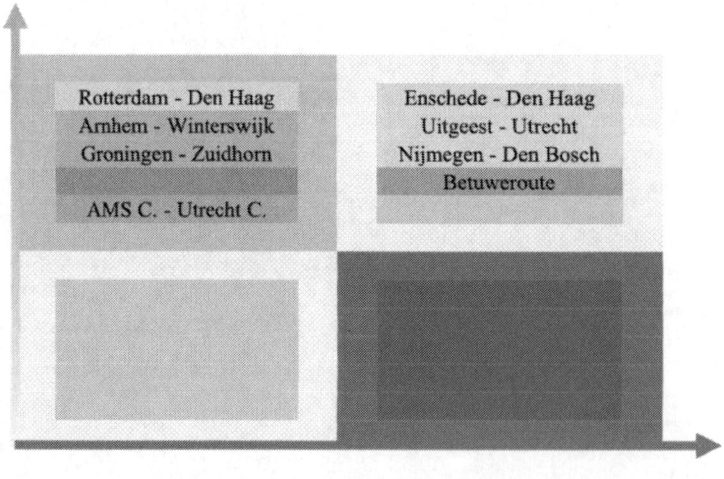

Figure. 6. Framework results for GoA 4

In GoA 4, there is an additional flexibility gain and staff costs could be further reduced. These additional gains are likely to compensate the costs in regional and relatively short main

/ freight corridors. On the other rail corridors, there is again doubt if the benefits from implementing ATO could compensate for the costs.

12 Discussion

The objective of the framework was to analyse rail corridors and determine the expediency of ATO. However, rail corridors are part of the railway network. Therefore, it is important to understand the impact of ATO over ETCS on the railway network considering the suggestions given by the framework.

The current corridor segmenting in the Netherlands has resulted in dependencies between the corridors. Therefore, in case ATO is decided on a corridor (e.g. Amsterdam Central – Utrecht Central), there would be transitions between manual and automated operation. These transitions could result in the following challenges:

- Information Presented on DMI
- Human Factor
- Planning and Scheduling for Train Driver

13 Framework validation

13.1 Review meetings

At the end of each stage, design review meetings were held with experts. In these review meetings, railway experts and university scholars were present. Methods and results were discussed and if approved, next stage activities were carried out.

13.2 Applicability analysis

In order to determine the accuracy and acceptability of the designed framework, an applicability analysis is done. The framework was applied to two currently automated urban rail systems. This was due to the similarity of the architecture and operational principles of ATO in heavy rail and the urban rail system.

The chosen corridors are Canada line, which is operating with GoA 4, and Amsterdam Noord – Zuid line that is capable of operating with GoA 4.

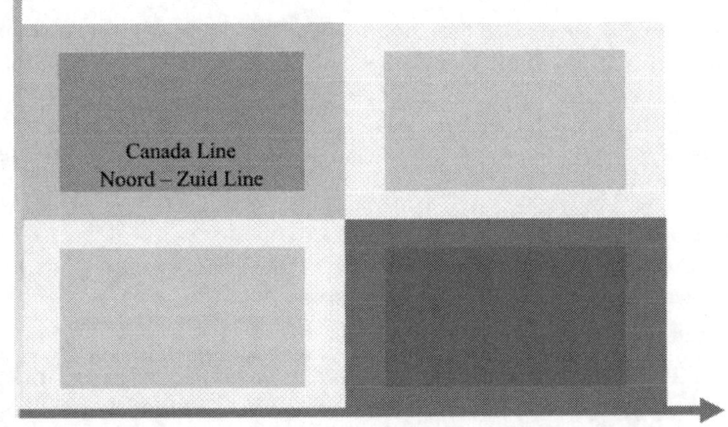

Figure. 7. Applicability results for GoA 4

The framework supports the decisions to implement GoA 4 for Canada line and Noord-Zuid line. Consequently, it can be concluded that the framework returns reliable results.

14 Recommendations

The recommendations for future work include:
1. A return of investment analysis along with a quantitative approach is suggested for a detailed study and determining the break-even points.
2. The impact of transitions between manual and automated operation should be further studied in order to determine and manage the challenges which could occur as a result of transitions.

15 Conclusion

In conclusion, the main objective of this research was to develop a tool which could suggest expediency of ATO over ETCS Level 2 in the heavy rail for the Dutch railway. The methodology and approach taken during this research resulted in a framework which fulfilled the objective. The following results were obtained after applying the framework to corridors in the Dutch railway:

- The framework suggests that GoA 2 may not be suitable for the Dutch railway. However, GoA2 is an entry door towards automatic operation and ensures smooth migration towards higher GoA,
- A positive business case for a few corridors in the Netherlands could be defined for GoA3. These corridors include relatively short main corridors which are segregated and regional corridors which have many stations.
- Regional corridors and relatively short main corridors could be suitable for GoA 4.
- Corridors with dense traffic which are relatively long may not be suitable for automated operation.

However, considering the current corridor segmenting and dependencies between the corridors, additional challenges would result from implementing automation. These challenges are due to the transitions between manual and automated operation and could even limit the benefits gained from automation.

Acknowledgment

The authors very much appreciate the support of Henk Barten, Aspectmanager Toelating Veilige Integratie at NS, and Mark Van Hesse, Sr Consultant Railway Signalling & Safety at Mott MacDonald, for their support and providing valuable insight.

The authors also thank the experts who were involved in the survey. Without their passionate participation and input, the survey could not have been successfully conducted.

References

1. https://projects.shift2rail.org/s2r_projects.aspx
2. ETCS System Description, RSSB (2010)
3. IEC, IEC 62290 - Railway applications - Urban guided transport management and command/control systems, Part 1: System principles and fundamental concepts (2014)
4. Railmap 4.0, ERTMS (2019)

5. R. Hayes, S. Wheelwright, *Restoring Our Competitive Edge: Competing Through Manufacturing* (1984)
6. Department of Defence, *Standard Practice System Safety* (2000)
7. Official Journal of the European Union, Directive 2016/798, (2016)
8. M. Abril, F. Barber, L. Ingolotti, M.A. Salido, P. Tormos, A. Lova, An Assessment of Railway Capacity, Transportation Research Part E: Logistics and Transportation Review, **44**, (2008)
9. C. Fombrun, Reputation Management: A Framework for Measurement and Valuation (2005)
10. J. Moses, Flexibility and Its Relation to Complexity and Architecture in Complex Systems Design and Management, Springer (2010)
11. A. Froger, O. Guyon, E. Pinson, A set packing approach for scheduling passenger train drivers: the French experience (2015)

 https://doi.org/10.1051/matecconf/202031401005

Appendices

Appendices A

Category	KPI \ Parameters	Corridor Characteristics																ATO		
		Avg. Distance between Stations	No. Of Stations	Max. Track Speed	Number of Junctions	Suicide history	Corridor Length	Track segregation	Average Track Adhesion Conditions	No. of Level Crossings	Available Infrastructure Capacity	Infrastructure Conditions (good, poor)	Traffic Heterogeneity	Current number of Trains per hour	Train Type (passenger, freight)	Signalling System (ETCS)	No. of passengers	GoA 2	GoA 3	GoA 4
	1. Safety																			
	- Number of SPADs																			
	- Number of over-speeding																			
	- Number of collision with object or person near platform																			
	- Number of collision with object or person on the track																			
	- Number collision with road traffic at Level Crossings																			
	2. Availability																			
Performance	- Time to remove a faulty train from the track after a failure																			
	3. Punctuality																			
	-Variations from dwell time																			
	- Variations from running time																			
	4. Capacity of corridor																			
	- Difference between theoretical and practical capacity																			
	5. Quality of Service																			
	- Percentage of inconvenience acceleration or braking																			
	6. Reliability																			
	- Number of disruptions due to technical failures																			

	Parameters	Corridor Characteristics																	ATO		
Category	KPI	Avg. Distance between Stations	No. Of Stations	Max. Track Speed	Number of Junctions	Suicide history	Corridor Length	Track segregation	Average Track Adhesion Conditions	No. of Level Crossings	Available Infrastructure Capacity	Infrastructure Conditions (good, poor)	Traffic Heterogeneity	Current number of Trains per hour	Train Type (passenger, freight)	Signalling System (ETCS)	No. of passengers	GoA 2	GoA 3	GoA 4
	7. Reputation																				
	Innovation																				
	- Percentage of automated driving tasks																				
	Human Factors																				
	- Public acceptance																				
	- Driver's resistance																				
Time to Service	**8. Flexibility**																				
	- Logistic complexity related to driver's planning																				
	9. Migration																				
	- Lead time to implement / adopt ATO																				
Financial	**10. Operational Expenses**																				
	- Energy usage per km per train																				
	- Staff cost per train																				
	11. Investment Expenses																				
	- Trackside installation expenses																				
	- On-board installation expenses																				
	12. Maintenance Expenses																				

		Corridor Characteristics																	ATO		
Category	Parameters / KPI	Avg. Distance between Stations	No. Of Stations	Max. Track Speed	Number of Junctions	Suicide history	Corridor Length	Track segregation	Average Track Adhesion Conditions	No. of Level Crossings	Available Infrastructure Capacity	Infrastructure Conditions (good, poor)	Traffic Heterogeneity	Current number of Trains per hour	Train Type (passenger, freight)	Signalling System (ETCS)	No. of passengers	GoA 2	GoA 3	GoA 4
	- Maintenance expenses of Infrastructure Manager																			
	- Maintenance expenses of Railway Operator																			

Legend	
Corridor	
	No Relation
•	Relation
ATO	
↗↗	Strong Increase (more than 50%)
↗	Weak Increase (less than 50%)
→	No Change
?	Unknown
↘	Weak Decrease (less than 50%)
↘↘	Strong Decrease (more than 50%)
< 50%	Weak
> 50%	Strong

Appendix B – Survey Participants

Role	Organization
Head of Quality Management	Railway Undertaking
Senior Rail Manager	Consultancy Firm
Consultant Railway Signalling & Safety	Consultancy Firm
Consultant Transportation	Consultancy Firm
Manager systeemintegratie	Infrastructure Manager
Programmemanager Smart & Green Mobility and Public Transport	Government Official
Manager Safety	Railway Undertaking
Programmamanager ATO	Railway Undertaking
ATO Project Managar	Signalling System Manufacturer
Advisor Innovation and Development	Infrastructure Manager
Consultant Railway Safety & Signalling	Consultancy Firm
Implementatie manager	Railway Undertaking
Consultant Integration Management	Consultancy Firm
Project Manager for ATO	Infrastructure Manager
ATO Project Managar	Signalling System Manufacturer
Consultant Railway Signalling & Safety	Consultancy Firm
Consultant Transportation	Consultancy Firm

Brownout Simulations of Model-Rotors In Ground Effect

F. Rovere[1,*], *G.N.* Barakos[1,†], and *R.* Steijl[1,‡]
[1]CFD Laboratory School of Engineering, University of Glasgow

Abstract. In this work computational fluid dynamics is used to validate experimental results for a two-bladed small rotor In Ground Effect conditions. The paper focuses on the evaluation and prediction of the rotor outwash generated in ground effect. Time-averaged outflow velocities are compared with experimental results, and the simulated flow field is used for safety studies using the PAXman model and particle tracking methods. The aircraft weights have been studied, evaluating scaling factors to define how helicopter weight can affect the outflow forces and the particle paths. Results show how the wake generated by heavier helicopters can lead to stronger forces on ground personnel and push the particles farther away from the rotor.

Nomenclature

Acronyms

CFD Computational Fluid Dynamics

DVE Degraded Visual Environment

IGE In Ground Effect

MTOW Maximum Take-Off Weight

MUSCL Monotone Upstream Centered Schemes for Conservations Laws OGE Out of Ground Effect

[*] PhD Student - federico.rovere@glasgow.ac.uk
[†] Professor - george.barakos@glasgow.ac.uk
[‡] Senior Lecturer - rene.steijl@glasgow.ac.uk

PIV Particle Image Velocimetry

Greek

N Kinematic viscosity m2/s

Ψ Local azimuth angle, deg.

ρ Density, kg/m3

θ_{75} Collective pitch at three-quarter radius, deg.

Latin

u Flow field velocity vector, m/s

$\mathbf{u_p}$ Particle velocity vector, m/s

a Speed of sound, m/s

$\mathbf{a}_p$ Particle acceleration, m/s^2

B Ballistic coefficient, kg/m^2

c Blade chord, m

C_D Drag coefficient

C_Q Rotor torque coefficient, $C_Q=Q/(0.5\ \varrho\ V^2_{tip}\ S_{rotor}\ R)$

C_T Rotor thrust coefficient, $C_T=T/(0.5\ \varrho\ V^2_{tip}\ S_{rotor})$

d_p Rotor diameter, m

g Gravity acceleration, m/s^2

h_{PAXman} PAXman height scaled, m

M Mach number, $M=V_{tip}/a$

m_p Mass particle, kg

N_b Number of blades

Q Rotor torque, N · m

R Rotor radius, m

r Radial coordinate along blade span, m

Re Reynolds number, $Re = V_{tip}\, c/\nu$

S_p Frontal particle area, m^2

S_{rotor} Rotor disk area, m^2

T Rotor thrust, N

U Velocity x-component, m/s

$\mathbf{u}^*$ Friction velocity, m/s

V Velocity y-component, m/s

v_i Hover induced velocity,

v_{max} Highest value of radial velocity

$V_{rad} = U\cos(\Psi) + V\sin(\Psi)$, m/s

Super and sub scripts

fs Full scale

p Particle

ss Small scale

tip Blade tip value

∞ Freestream value

1 Introduction

The flow fields generated by rotors operating In Ground Effect (IGE) are complex and unsteady and may also interact with the ground plane. The result of this interaction is the transition of the rotor induced flow from vertical (downwash) to radial flow (outwash). This flow field can be the source of risks for ground personnel, equipment and landscape due to the forces generated by high outflow velocities. Furthermore, in case of brownout or whiteout the flow field can interact with a loose sediment bed, uplifting particles all around the aircraft. The prediction of the outwash is fundamental for safety of helicopter IGE operations. The downwash typically affects activities directly under the aircraft (like search and rescue operations), while the outwash impacts on the surrounding environment like people, equipment and structures during landing and take-off. As already mentioned, brownout and whiteout may occur when a rotorcraft is operating in ground proximity and they are due to the interaction of rotor wake with the particles of a loose sediment bed such as sand, snow etc. This interaction can eventually cause the uplift of particles from the ground and their entrainment into the air flow. When operating in desert areas or in snowy regions, the number of particles entrained can become extremely high creating a cloud around the rotorcraft. The main effects of this cloud of particles moving around the aircraft are on the pilot's visual environment and on the rotorcraft structure and equipment. The lack of pilot visibility defines brownout and whiteout as Degraded Visual Environment (DVE) conditions. In recent years, efforts have been made to help pilots in these situations, developing sensors and advanced cockpit displays. Dynamic rollover and collisions with objects are common accidents due to the lack of visibility [1]. In [2] NATO statistics are given about mishaps due to brownout or whiteout: US Air force lost 30 special operation aircraft and 60 crew members lost their lives during landing in desert environments since 1990. In the same report authors specify that brownout cost to US services $100M/yr. Other NATO members experienced similar statistics. UK had 24 brownout mishaps in the period 2005-2009. The German defence force had more than 30 accidents due to dust and snow. In [3], the authors refer that the occurrence of brownout is the most common cause of human factor mishaps during military operations. In the present work, computational fluid dynamics (CFD) analysis of a micro two-bladed rotor operating IGE is compared with experimental results. The test case simulated was experimentally investigated at the University of Maryland by Lee et al. [4]. The flow field data was obtained by 2D Particle Image Velocimetry (PIV), while the rotor performance was measured by a micro mass balance. Once the CFD analysis was complete, the outflows produced were used to evaluate the forces produced on ground personnel, properly scaled, and using the PAXman model, and particle tracking. To obtain realistic full-scale scenarios, a scaling factor has been applied to velocities using three different aircraft, categorized in terms of weight. After this, safe zones are defined where it was possible to consider the outflow forces and the presence of the particles as negligible. Existing safety separation criteria can then be tested. A possible separation distance is suggested by FAA for wake encounters [5] [6]. In that case, a distance of 3 rotor diameters is suggested to allow the dissipation of the wake generated by a rotor in hover or taxiing. Investigations are conducted to verify if the same distance can be considered safe for forces due to the outflow acting on ground personnel and for the presence of particles in the area that can spoil the pilot view of a near aircraft or hit people operating inside the 3 rotor diameters area. All CFD simulations have been performed using the HMB3 (Helicopter Multi-Block) CFD solver of Glasgow University.

2. CFD solver

The HMB3 (Helicopter Multi-Block) [7] [8] is the solver used for CFD calculations in this work. It solves the Unsteady Reynolds Averaged Navier-Stokes equations (URANS) in integral form with ALE formulation (Arbitrary Lagrangian Eulerian) formulation for time-dependent domains (moving boundaries). URANS equations are discretised using a cell-centred finite volume approach on a multiblock structured grid. HMB uses the Osher [9] and Roe [10] approximate Riemann solvers to evaluate the convective fluxes, the viscous terms are discretized using second order central differencing. Third order accuracy in space is provided by The Monotone Upstream Centred Schemes for Conservation Laws (MUSCL) [11]. An implicit dual time stepping methods is employed to perform the temporal integration. Oversets grids (used in this work) [12] and sliding plane [13] methods are available in HMB to allow for the relative motion between mesh components, representing ground and rotor blades.

2.1 PAXman model

The PAXman model was originally developed for military personnel and it is used to compute wind forces on people operating in proximity to the rotor. It is based on the projection of a crouching 6ft tall person immersed in the outwash. The details of the geometry of the PAXman model are shown in figure 1, as reported in [14] and [15]. To be comparable with the small rotor size, the PAXman height is scaled with the rotor radius for three different full-scale rotors, chosen to represent three different weight categories: light, medium and heavy. Their technical data are listed in table 2.

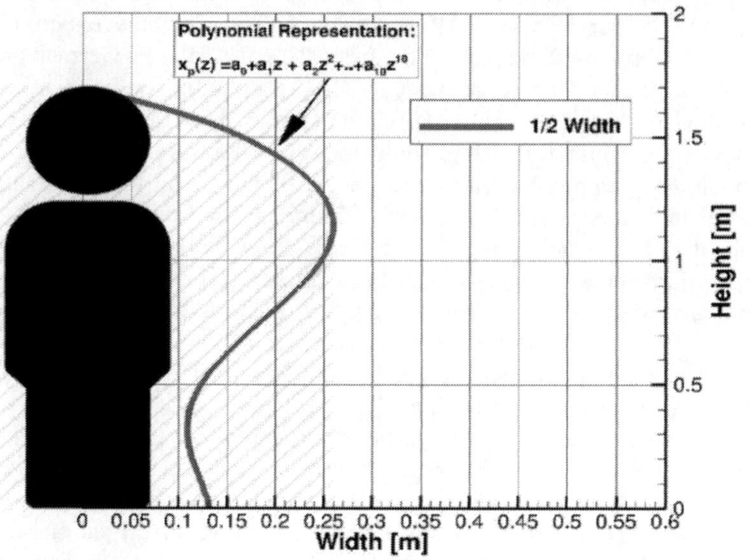

Figure 1: PAXman model [15]

Table 1: Polynomial coefficients used in figure [1]

a0	4.30939e-01
a1	-4.63972e-02
a2	-1.39649e-01
a3	1.37545e-01
a4	1.37545e-01
a5	-5.49253e-04
a6	2.21653e-04
a7	-4.18444e-05
a8	1.45194e-05
a9	-7.80009e-08
a10	-1.89822e-07

Table 2 Helicopters technical data [16]

Category	MTOW (kg)	R (m)	$V_{tip\ (m/s)}$	C_T
Light	3000	5.5	220	0.009
Medium	8000	8.1	216	0.015
Heavy	11000	8.1	220	0.0176

The PAXman height has been scaled with the rotor radius, giving of $h_{PAXman} = 0.3R$ for the lighter aircraft, while for the medium and the heavy cases the result is the same, due to the similar size of the rotor radius, which lead to h_{PAXman}= 0.2R. Using the reference area expressed by the polynomial representation of figure 1, the distribution of the force produced by the outwash on the ground personnel is computed as proposed in [14]. To obtain comparable forces with the safety criteria, velocities are scaled using the blade tip velocity, listed in table 2. For this first scaling is necessary, to obtain velocities expressed in m/s, that can be used, later, to compute the PAXman forces in N. However, a second scaling is necessary to take into account the difference in C_T of the three different cases. A common reference value for outflow velocities is the dimensionless hover induced velocity λ_i , expressed as $\lambda_i=\sqrt{(C_T)}/2$. The scaled rotor thrust coefficient obtained by the simulation of the micro rotor is $C_T^{ss} = 0.03$, while the C_T^{fs} can be obtained by the data, considering the aircraft in hover flight, and thrust equal to the weight, that in this case is considered the maximum at take-off. W_{MTOW} ,V_{tip} , and S_{rotor} or are specified in table 2, while for the acceleration of gravity and for the air density the following values have been assumed: $g = 9.81 m/s^2$ and ρ_{air}= $1.225 kg/m^3$

$$C_T = \frac{2W_{MTOW}}{0.5\rho V_{tip}^2 S_{rotor}} \quad (1)$$

It is possible, taking into account the effect of the different thrust coefficient to scale the velocities using the ratio of hover induced velocity between full-scale and small-scale cases. The scaling factor obtained is $\sqrt{C_T^{SS}/C_T^{fs}}$. This way, it is possible to estimate the outflow velocities generated by a full-scale rotor operating at the same high values of thrust coefficient of the scaled rotor. According to [14] and [15] the caution zone starts when the force acting on the PAXman is more than 80 lbf (335 N), and the hazard zone is defined after 115 lbf (510 N). The distribution of the force over the body is calculated as: $f_{PAXman} = 0.5\rho_{air}V2_{rad}x$ where ρ_{air} is density of the air, V_{rad} the radial velocity and x is the horizontal coordinate of the PAXman model. The total force is the integral of the distribution of the force over the height of the PAXman model.

$$F_{PAXman} = \int_{h_{PAXman}} f_{PAXman}dx \quad (2)$$

Using this model, it is possible to calculate the force distribution acting on a human body at a specific radial station.

2.2 Particle tracking

Brownout and whiteout are due to the presence in the flow field of particles. The former involves sand, and the latter snow. However, other kinds of particles can be involved such as rain, ice and even small rocks. To properly simulate the behaviour of them in the flow field it is necessary model their motion. There are basically two approaches for the numerical simulation of dispersed phases, and they can be categorized into Lagrangian tracking or Eulerian modelling approaches. In the Lagrangian approach, the particles (or parcels of particles) are tracked through the field and the local cloud properties are defined by their properties as they pass the point in the field. For methods that involve this approach the motion of the particles is tracked solving the Newton's second law. Important works in Lagrangian frame of reference are [17] [18]. In the case of Eulerian methods, the properties of the particles are assumed to be continuous within the field. Thus, differential conservation equations are written, discretised, and the solution of these gives the properties of the cloud. [19] [20]. For particle tracking, an in-house software has been used. The particles are driven by the flow field velocities and their positions in time are obtained by integrating their equations of motion. The integration method used is a fourth order Runge-Kutta, and the equation for particle tracking acceleration is.

$$\boldsymbol{a}_p = \frac{0.5\rho_{air}(u-u_p)\|u-u_p\|}{B} - \mathbf{g} \tag{3}$$

Where $\mathbf{a}_p$ is the acceleration of the particle, $\mathbf{u}_p$ is its velocity, $\mathbf{u}$ is the velocity of the flow field in the position of the particle and B is the ballistic coefficient, $B = m_p/(S_p - C_D)$. Here m_p is the particle mass,S_p is the particle frontal area (particles are assumed spherical) and C_D is the particle drag coefficient, and finally **g** is acceleration of gravity. The particle proprieties used in this work are listed in table 3, they reflect the size and the density of particles used to simulate brownout in experimental and computational works [21] [22] [23].

Table 3 Properties of particles used in this work.

ρ_p (kg/m^3)	d_p (µm)	C_D	B
2650	9	1.048	0.03

3. Computations and safety considerations

Previous works show the full validation of the current methods [24]. Here a brief summary of the work is presented. IGE configuration the rotor was modelled using two overset grids, and the ground was modelled imposing no-slip and no-penetration conditions. The full rotor domain was computed as unsteady. Comparisons in terms of performance are given in figure 2 and show small differences with of the experiments of [4]. The thrust coefficient shows results very close to the result proposed by Lee et al. [4]. The increment of thrust coefficient IGE is about 12% with respect to the OGE case. The result of the power coefficient is slightly higher C_Q^{IGE} / C_Q^{OGE} =1.02. However, it can be considered that the power between OGE and IGE is fairy than what expected constant, with a small increment in the thrust coefficient, as expected. Similar results in terms of C_Q^{IGE} / C_Q^{OGE} have been obtained by Karla et al. in [25] for the same test case.

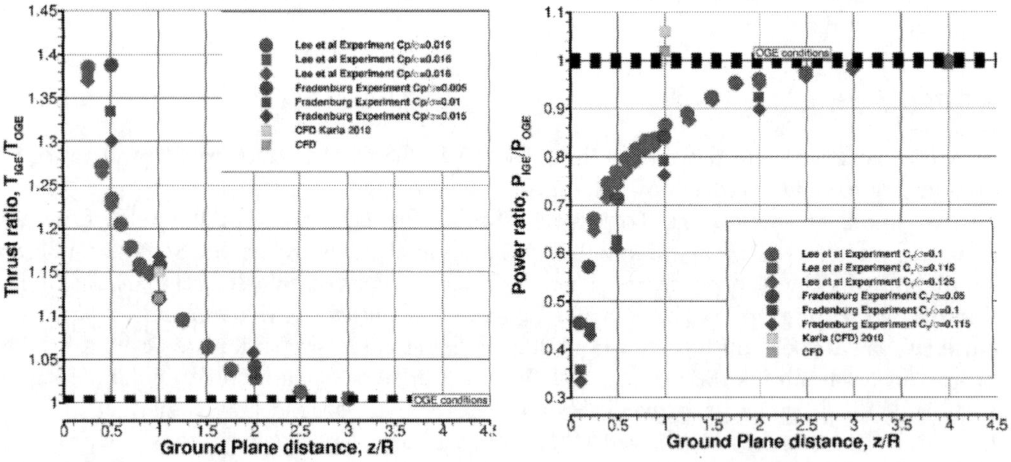

(a) C_T^{IGE} / C_T^{OGE} (b) C_Q^{IGE} / C_Q^{OGE}

Figure. 2: Experimental and CFD results for the trust (left) and power (right) ratios IGE. The rotor was operating at θ_{75} = 12deg, Re_{tip} = 35000 and M_{tip} = 0.08. Experimental data taken from [4] and additional CFD data from [25].

In figure 3, the rotor wake is presented as (b) iso-surfaces of Q-criterion and (a) contours of vorticity magnitude. The Q criterion is related to the velocity gradient tensor, ant its iso-surfaces are a good indicator of how vortical flow may be [26]. As expected, the wake, reaches the ground and expands. The presence of tip vortices and vortex sheets are also expected

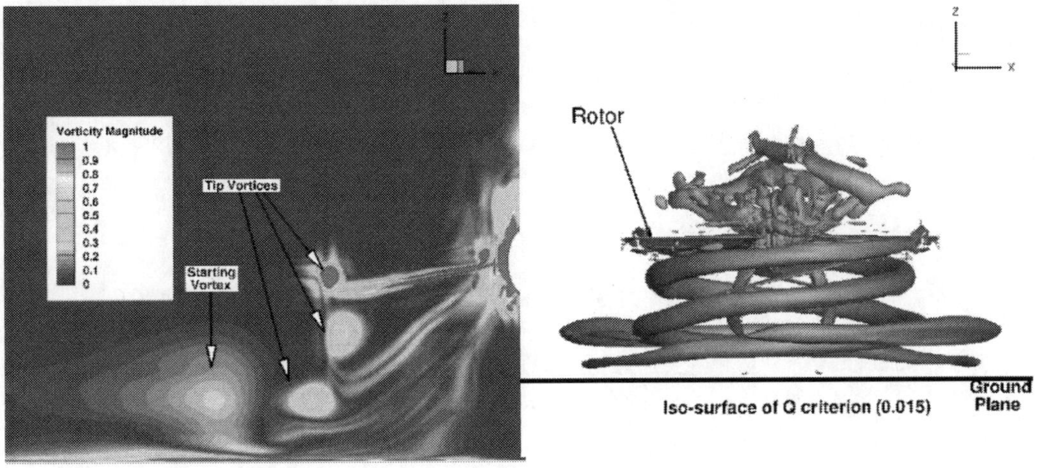

(a) Vorticity (b) Q-criterion isosurfaces

Figure. 3 Wake Visualization, for the IGE case. The rotor operating at θ_{75} = 12deg, Re_{tip} = 35000, M_{tip} = 0.08, h/R = 1 and C_T= 0.03.

In figure 4 the time-averaged radial velocities are shown in comparison with experiments. The time-averaged outflow has been scaled using two different velocities. The left side of figure 4 shows results scaled with the reference hover induced velocity (as proposed by Lee et al. [4]), while on the right side of figure 4 the same results are scaled as a jet, using the peak radial velocity v_{max}. Finally, the dashed line represents the PAXman model height. The time-averaged velocities show good agreement with experiments in terms of maximum radial velocity and outflow distribution for both scaling methods, figure 4 (a) (b), (c) and (d), near the rotor. For higher radial positions, the peak velocity is underestimated, and its position is higher with respect to the ground, with some differences over the height distribution shown in figure 4 (e). This could be due to the few rotor revolutions performed in the simulation.

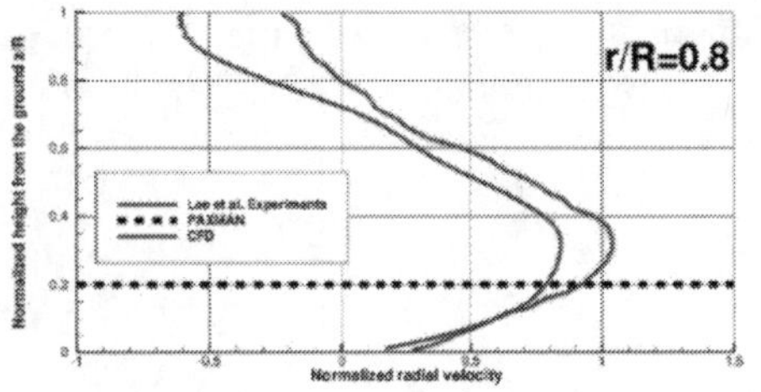

(a) V_{rad} / λ_i at r/R=0.8

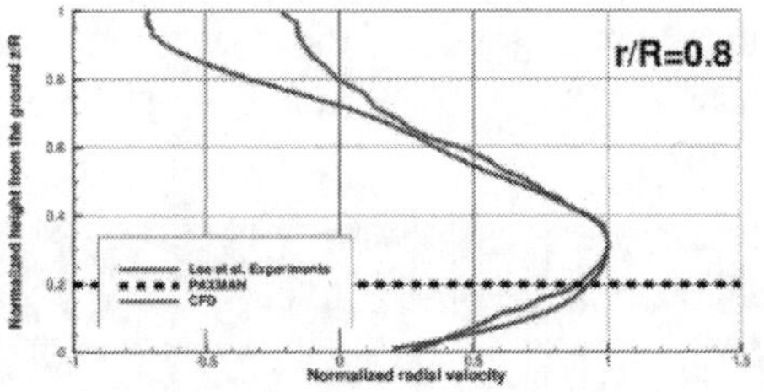

(b) V_{rad}/V_{max} at r/R=0.8

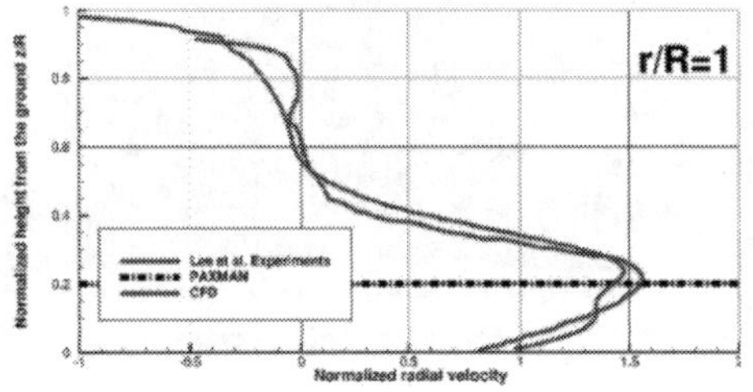

(c) V_{rad} / λ_i at r/R=1.0

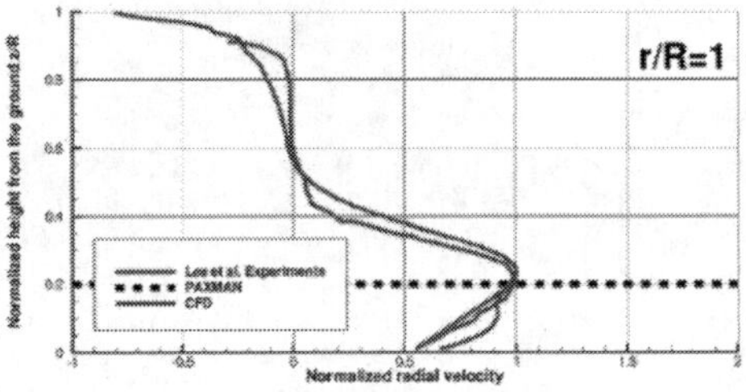

(d) V_{rad}/V_{max} at r/R=1.0

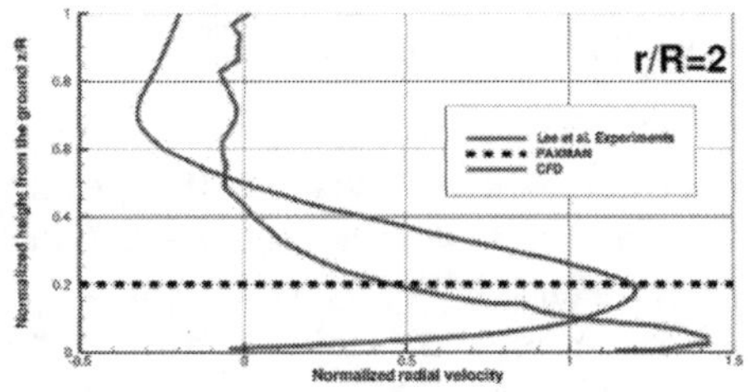

(e) V_{rad} / λ_i at r/R=2.0

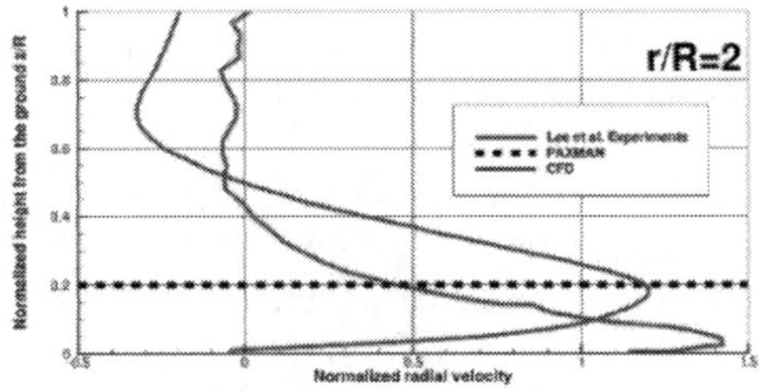

(f) V_{rad}/V_{max} at r/R=2.0

Figure. 4 Experimental and CFD time-averaged outflow velocity profiles at different radial positions. The rotor was operating θ_{75}= 12deg, Re_{tip} = 35000 M_{tip} = 0.08 and C_T = 0.03. Left column results are scaled with induced hover velocity, while the right column results are scaled with jet-scaled velocity. Experimental data taken from [4].

Figure 5 (a) shows the total forces acting on a scaled human body at different distances by the outflow produced by the rotor. As mentioned before, the forces have been computed scaling the velocities with the V tip of full-scale aircraft, to obtain more realistic values and compare them with the safety threshold suggested for civilian and military operations. The forces for all three aircraft reach their peaks at a radial distance between 1.25R and 1.5R, but they weaken for larger radial stations. This result is due to the outflow distribution at different radial stations, with the higher velocities occurring around 1.5R. As expected, the strongest force is produced by the aircraft with higher C_T, which is the heaviest helicopter taken into account. The outflow forces produced by the lightest helicopter are small in comparison to the other two helicopters and also with respect to the safety thresholds suggested by the PAXman method. For the other two cases, the forces exceed the caution limit, but do not reach the danger zone threshold. This result is in agreement with other works, such as in [15] where the PAXman forces of a V-22 have been computed starting from an experimental outflow survey. This reference involved several configurations of disk loading and it has been shown that higher values of disk loading result in forces that reached a peak of 600 N. In all three cases, after 2R, the force drops quickly. This is due to the outflow velocities that drops away from the rotor. After 3R the forces are almost negligible based on the PAXman model. Figure 5 (b), (c) and (d) show the distributions of the forces over the PAXman height at different radial stations for the three helicopters. For all three cases, at early distance from the rotor the force is strongest at a specific PAXman height (corresponding to the PAXman chest). This result is due to the shape of the outwash distribution and the larger blocking area at the chest level. This result is in accordance with other experiments. In [14] the force peaks were around the middle of the human body, however the experiments performed by Silva et al., were full-scale. Figure 5 (b) shows the results for the light aircraft. Here the PAXman height is higher with respect to the other two cases. This is due to the smaller size of the rotor blade of the lightest helicopter. Furthermore, the forces at every radial distance are smaller respect to heavier cases. This is in agreement with what is shown in figure 5 (a), where the lightest produces weaker forces in comparison to the medium and the heavy.

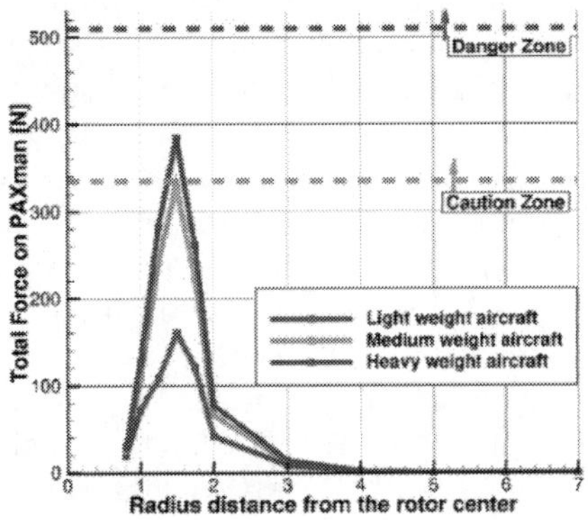

(a) Total radial force F_{PAXman} for the three helicopters

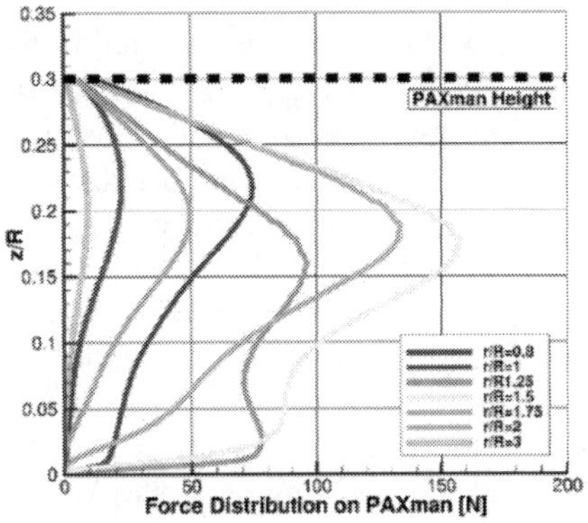

(b) Radial force distribution f_{PAXman} for light weight aircraft

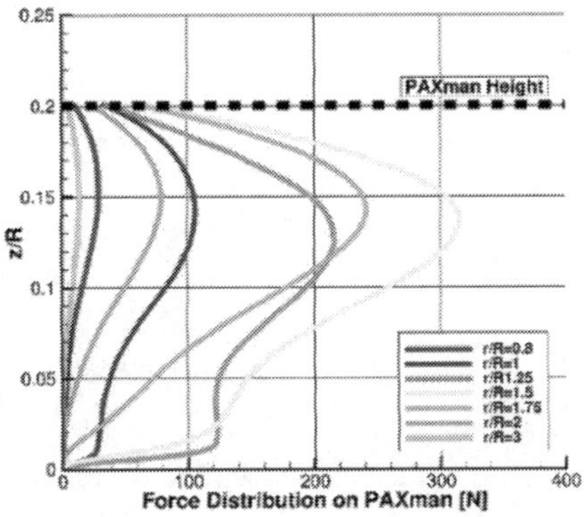

(c) Radial force distribution f_{PAXman} for medium weight aircraft

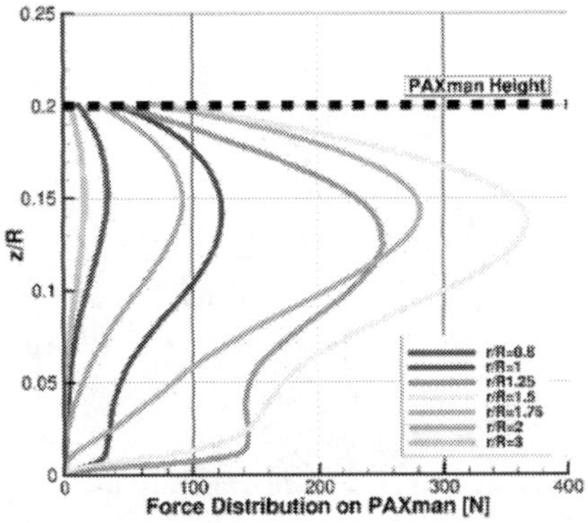

(d) Radial force distribution f_{PAXman} for heavy weight aircraft

Figure. 5 PAXman model forces calculated using the employed micro-rotor, scaled to V tip m/s. The micro rotor was operating at θ_{75}= 12deg, Re_{tip} = 35000, M_{tip} = 0.08, h/R = 1 and C_T = 0.03. Scaling V_{tip} and C_T for full scale rotor are listed in table 2.

In [24] the computed shear stress on the ground of the domain allowed to calculate the particle velocity uplift threshold. It was shown that the peak of the uplift ratio u_t/u^*_t is in proximity of 1 R distance from the rotor. Using this information, it was possible to seed the ground with the particles that are more likely to be affected by the flow field. Figure 9 shows the evolution of the particles with time for two weight helicopter cases taken into account, with the rotor at two different heights above the ground h/R=1 and h/R=0.5. Different seeding positions have been defined between 1R and 1.5R in proximity of the ground level. Bombardment effect has not been considered for these simulations. These simulations involve about 100 revolutions, for a total time of 30 seconds. The flow field generated by the model rotor was not able to lift particles. As for the PAXman force computations, to obtain more realistic results the flow field has been scaled using a possible operational V tip for light and heavy weight helicopters considered in table 2, and with the C_T scaling factor. When the rotor is operating at h/R=1, the particles are uplifted by the flow field, and then move away from the rotor, following the radial direction. Particles are driven by the outflow that pushes them away from the rotor. However, depending on the strength of the outflow they reach different positions. Heavy weight helicopters have a stronger outflow, and in this case, particles can reach a maximum radial distance of 8.5R, and a maximum height above the ground of 1R. Once the particles reach the maximum altitude value, they fall again on the ground. The light weight helicopter case, shows lower values for maximum h/R and r/R (0.75R and 6.5R), however particles go further, the 3D separation criteria for wake encounters. On the other hand, when the helicopter is operating at h/R=0.5 particles show a different behaviour. Initially particles are uplifted from the ground, then two main branches spread, following different paths. Some of them, are re-ingested by the rotor, reaching the highest distance from the ground (around 2R for heavy weight test case and 1.75R for light weight). These particles can be dangerous for the crew and the aircraft. The rest of the particles keep following the radial direction far from the rotor and fall again on the ground (almost 7R for heavy weight helicopter and more than 6R for light). Particles that move away from the rotor can be dangerous for ground personnel and equipment, while the recirculation of the particles creates risks for the helicopter and the crew itself. During the re-ingesting phase particles can hit blade and fuselage and they create the cloud that degrades the visual of the pilot, creating a dangerous DVE condition.

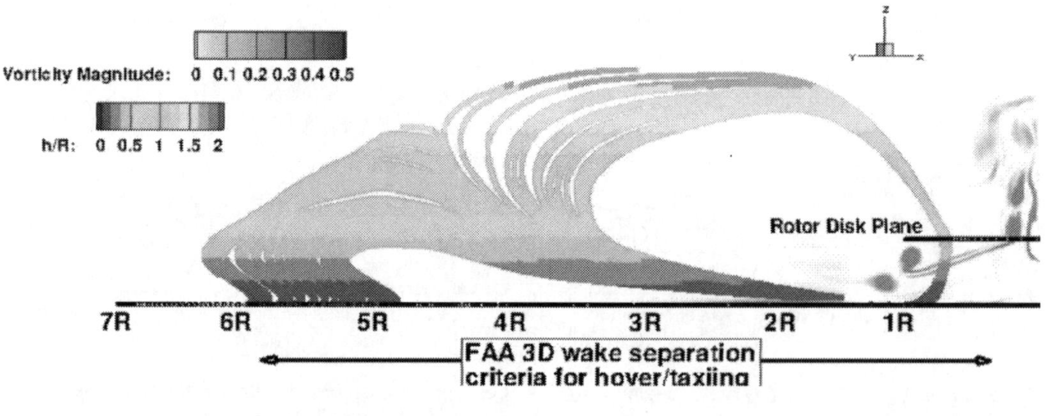

Figure. 6 Particle tracking, using velocities scaled with the light weight aircraft, tip velocity (V tip = 220 m/s) and thrust coefficient for hovering at MTOW : C_T= 0.009. The rotor was operating at θ_{75}= 12deg, Re_{tip} = 35000, M_{tip} = 0.08, h/R = 0.5 and C_T= 0.035. Contours correspond to the magnitude vorticity and the FAA 3 diameters separation distance is shown for comparison [5].

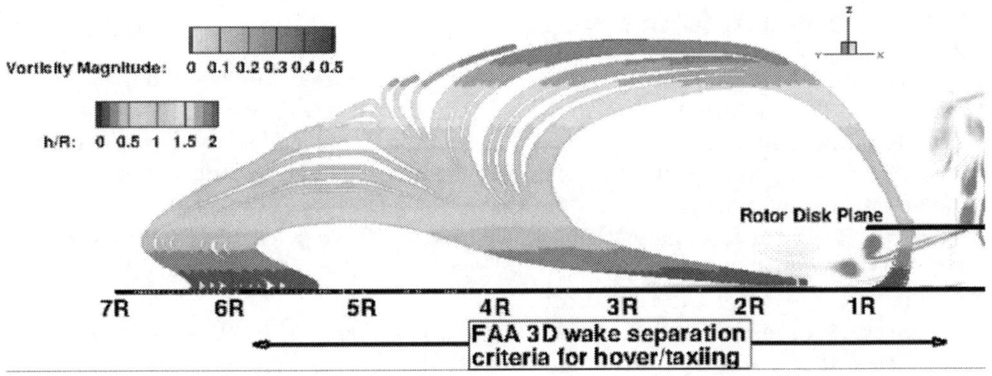

Figure. 7 Particle tracking, using velocities scaled with the heavy weight aircraft, tip velocity (V tip = 220 m/s) and thrust coefficient for hovering at MTOW : $C_T = 0.017$. The rotor was operating at θ_{75}= 12deg, $Re_{tip} = 35000$, $M_{tip} = 0.08$, $h/R = 0.5$ and $C_T = 0.035$. Contours correspond to the magnitude vorticity and the FAA 3 diameters separation distance is shown for comparison [5].

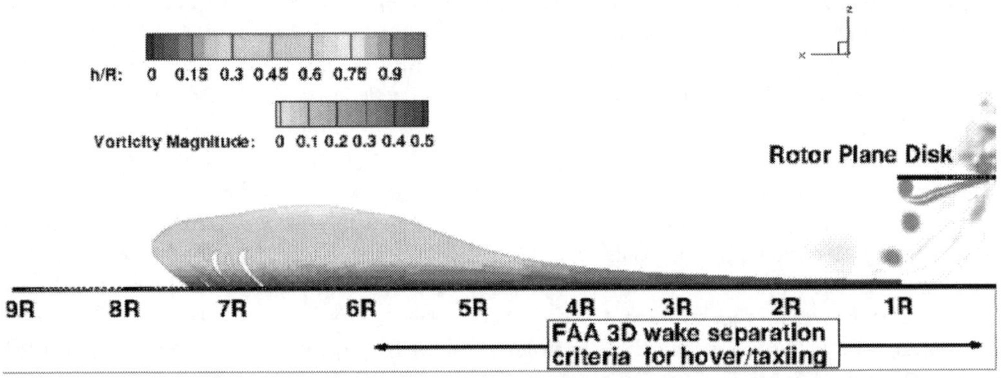

Figure 8. Particle tracking, using velocities scaled with the light weight aircraft, tip velocity (V tip = 220 m/s) and thrust coefficient for hovering at MTOW : $C_T = 0.009$. The rotor was operating at θ_{75} = 12deg, Re $_{tip}$ = 35000, Mtip = 0.08, h/R = 1 and $C_T = 0.03$. Contours correspond to the magnitude vorticity and the FAA 3 diameters separation distance is shown for comparison [5].

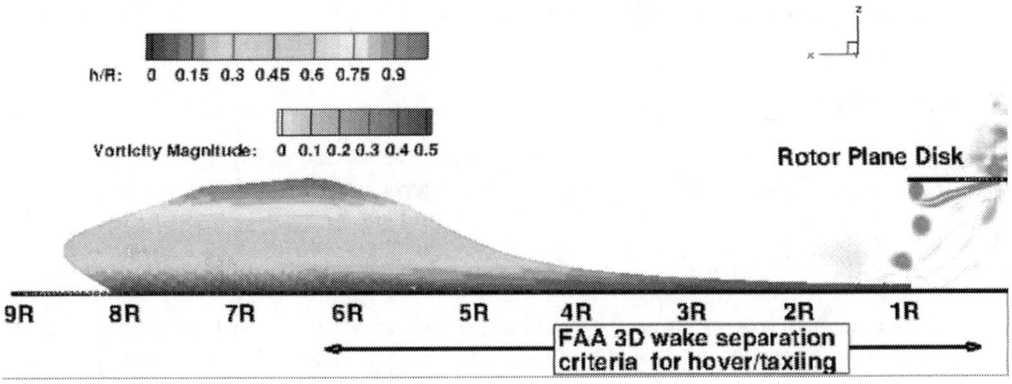

Figure. 9 Particle tracking, using velocities scaled with the heavy weight aircraft, tip velocity (V tip = 220 m/s) and thrust coefficient for hovering at MTOW : $C_T = 0.017$. The rotor was operating at θ_{75}=12deg Re_{tip} 35000, $M_{tip} = 0.08$, h/R = 1 and C_T= 0.03. Contours correspond to the magnitude vorticity and the FAA 3 diameters separation distance is shown for comparison [5].

4. Conclusions and future steps

CFD results demonstrate fairly good agreement with the experimental data in terms of rotor performance and flow field, even with some differences in the outflow predictions for larger radial stations. The differences can be important if used to evaluate safety regions near the helicopter. Results employing different scaling approaches show different agreements with experiments, and only scaling with the hover induced velocity gives a complete understanding of the outflow analysis in terms of safety. The normalized magnitude of the radial velocity peak in the outflow and its position, related to PAXman height, are fundamental to obtain correct force estimations. In general, safety considerations show that the distance criterion based on the PAXman forces, can be deducted by the outflow analysis. Outflow forces after a distance of 3R are low for the three helicopter cases considered. Furthermore, for the lightest helicopter, there is no risk due to the outflow forces for ground personnel. It is clear that the wake encounter criterion of 3D can be adopted for ground operation, if the presence of particles on the ground can be excluded. A more detailed analysis shows that the forces obtained for the medium and heavy aircraft are high enough to be dangerous for personnel and equipment in an area between 1R and 2R away from the rotor, but after 3R it is safer. The force distributions suggest that the chest is the part of the human body that is most influenced by the force due to the outflow, and this result can be part of the evaluation on definite safety regulations for people acting in proximity of the rotor. Particle tracking results show that particles can reach large distances away the rotor, exceed the limit of 3D. In general, it seems that the FAA limit for wake encounters cannot guarantee safety in presence of particles on the ground. It is also clear that to define a particle free zone , it is necessary to take into account the rotor operating conditions due to the strong influence of the disk loading and in general of the size of the aircraft on the particle paths. Particle paths are also strong influenced by the position of the rotor with respect to the ground. In this work, scaling factors are applied to a small-scale rotor to obtain full-scale PAXman forces and particle paths, however, the full physics of the brownout cannot be simulated in this way, due to the several factors. In general, the Reynolds number that is involved in the small and the full-scale scenario is dissimilar, leading to differences in the uplift phenomena and to a different evolution of the brownout cloud. This study can be a starting point for evaluating safe operational zones around a helicopter. A future step will consist of high detailed simulations to identify a proper safe zone from particles and compare it with other distance safety regulations like the 3 rotor diameters separation distance for wake encounters suggested in the Manual of Air Traffic Service [5] [6].

Acknowledgments

This project has received funding from the European Union's H2020 research and innovation programme under the Marie Sktodowska-Curie grant agreement No 721920

Bibliography

[1] D'Andrea, A., and Scorcelletti, F., "Enhanced numerical simulations of helicopter landing manoeuvres in brownout conditions," American Helicopter Society International 66th Annual Forum Proceedings, Phoenix, AZ, 2010.

[2] Taskgroup, R., "Rotary-wing brownout mitigation: technologies and training," Tech. Rep. RTO-TR-HFM-162, NATO Sci.Technol. Org., 2010.

[3] Mapes, P., Kent, R., and Wood, R., "DoD helicopter mishaps FY85-05: findings and recommendations," US Air Force, 2008.

[4] Lee, T. E., Leishman, J. G., and Ramasamy, M., "Fluid dynamics of interacting blade tip vortices with a ground plane," Journal of the American Helicopter Society, Vol. 55, No. 2, 2010, pp. 22005–22005.
[5] CAA, "CAP 490: Manual of Air Traffic Services Part 1," 2015.
[6] Garcia-Dorado, I., Aliaga, D. G., Bhalachandran, S., Schmid, P., and Niyogi, D., "Fast Weather Simulation for Inverse Procedural Design of 3D Urban Models," ACM Transactions on Graphics (TOG), Vol. 36, No. 2, 2017, p. 21.
[7] Steijl, R., Barakos, G., and Badcock, K., "A framework for CFD analysis of helicopter rotors in hover and forward flight," International journal for numerical methods in fluids, Vol. 51, No. 8, 2006, pp. 819–847.
[8] Lawson, S., Woodgate, M., Steijl, R., and Barakos, G., "High performance computing for challenging problems in computational fluid dynamics," Progress in Aerospace Sciences, Vol. 52, 2012, pp. 19–29.
[9] Osher, S., and Chakravarthy, S., "Upwind schemes and boundary conditions with applications to Euler equations in general geometries," Journal of Computational Physics, Vol. 50, No. 3, 1983, pp. 447–481.
[10] Roe, P. L., "Approximate Riemann solvers, parameter vectors, and difference schemes," Journal of computational physics, Vol. 43, No. 2, 1981, pp. 357–372.
[11] Van Leer, B., "Towards the ultimate conservative difference scheme. V. A second-order sequel to Godunov's method," Journal of computational Physics, Vol. 32, No. 1, 1979, pp. 101–136.
[12] Jarkowski, M., Woodgate, M., Rokicki, J., and Barakos, G., "Towards consistent hybrid overset mesh methods for rotorcraft CFD," 2011.
[13] Steijl, R., and Barakos, G., "Sliding mesh algorithm for CFD analysis of helicopter rotor–fuselage aerodynamics," International journal for numerical methods in fluids, Vol. 58, No. 5, 2008, pp. 527–549.
[14] Silva, M., and Riser, R., "CH-47D tandem rotor outwash survey," AHS 67th Annual Forum, 2011.
[15] Preston, J. R., Troutman, S., Keen, E., Silva, M., Whitman, N., Calvert, M., Cardamone, M., Moulton, M., and Ferguson, S. W.,"Rotorwash Operational Footprint Modeling," Tech. rep., Missile Research Development and Engineergin Center Redstone Arsenal Al Missile Guidance Directorate, 2014.
[16] "Vertipedia," https://vertipedia.vtol.org/, Accessed: 23-09-2019.
[17] Alfred, J., Celi, R., and Leishman, J. G., "Flight Path Optimization for Brownout Mitigation Using a High-Fidelity Simulation Model," Journal of the American Helicopter Society, Vol. 62, No. 3, 2017, pp. 1–15.
[18] Wachspress, D., Whitehouse, G., Keller, J., Yu, K., Gilmore, P., Dorsett, M., and McClure, K., "A high fidelity brownout model for real-time flight simulations and trainers," 2009, pp. 278–301.
[19] Phillips, C., Kim, H. W., and Brown, R. E., "The flow physics of helicopter brownout," 66th American Helicopter Society Forum: Rising to New Heights in Vertical Lift Technology, 2010.
[20] Ghosh, S., Lohry, M. W., and Rajagopalan, R. G., "Rotor configurational effect on rotorcraft brownout," 28th AIAA Applied Aerodynamics Conference, 2010, p. 4238.
[21] Syal, M., Govindarajan, B., and Leishman, J., "Mesoscale sediment tracking methodology to analyze brownout cloud developments," 66th Annual Forum of the American Helicopter Society, Phoenix, AZ, 2010, pp. 11–13.
[22] Kutz, B., Günther, T., Rumpf, A., and Kuhn, A., "Numerical examination of a model rotor in brownout conditions," 2014, pp. 2450–2461.
[23] Tanner, P. E., "Photogrammetric characterization of a brownout cloud," 2011.

[24] Rovere, F., Barakos, G., and Steijl, R., "CFD validation of a micro-rotor in grond effect," 45th European Rotorcraft Forum, CFAS, 2019.
[25] Kalra, T. S., Lakshminarayan, V. K., and Baeder, J. D., "CFD validation of micro hovering rotor in ground effect," American Helicopter Society 66th Annual Forum Proceedings, Citeseer, 2010.
[26] Hunt, J. C., Wray, A. A., and Moin, P., "Eddies, streams, and convergence zones in turbulent flows," 1988.

MATEC Web of Conferences **314**, 01007 (2020) https://doi.org/10.1051/matecconf/202031401007
ICSC-ISATECH 2019

Composite risk index: the new Safety Performance Indicator of risk exposure

Tamara Pejovic[1,*†]
[1]EUROCONTROL, Aviation Intelligence, Performance Review Services, Belgium

Abstract. Conceptually, all safety programmes desire accurate safety risk quantification in order to provide a meaningful expression of risk. As there are typically, multiple safety risks associated with a system or event, the quantification of total safety risk is a major challenge. One possible way to define and accept the total safety risk of any system is using the concept of a composite risk estimate. This paper represents development of the new safety performance indicator and overall methodology that could be used to measure the performance of European ATM systems as a whole and its individual entities. It describes the computation of the Composite Risk Index (CRI), logic behind it, its use (on the example EUROCONTROL Member States) and limitations and areas of potential improvement. CRI represents a cumulative risk value calculated aggregating all reported, assessed and severity classified key safety-related incidents to form an index. This measure of risk exposure is based on probability and severity that considers the human perception of equivalent risk. Overall idea behind CRI is that the performance of safety system can be analysed within three important broad categories: the quality of reporting system with reporting entity, measured risks within the system, and human perception of risk.

1 Introduction

Risk is the potential for mishaps or other adverse variation in the cost, schedule, or safety performance of ATM system. Safety risk therefore can be explained as the potential for mishaps that could result in injury, fatality, equipment or system damage or total loss.

Conceptually, all safety programmes desire accurate safety risk quantification in order to provide a meaningful expression of risk. As there are typically multiple safety risks associated with a system or event, the quantification of total safety risk is a major challenge.

One possible way to define the total safety risk of any system is using the concept of a composite risk estimate [1]. Current methods of obtaining this composite risk estimate use summing techniques to add the individual risks and produce a single number [1, 2]. This method seems natural, however, it is often difficult to determine particular occurrence probabilities (e.g. when historical information is of limited time series) or to quantify their

* Corresponding author: tamara.pejovic@eurocontrol.int
† Disclaimer: The views expressed in this paper are the authors' own and do not represent a policy or position of EUROCONTROL.

severity (e.g. when information in safety databases is missing). That makes the additive computation of risk difficult or impossible.

Moreover, although risk in general can be quantified, as it represents combination of probability and severity of specific occurrence happening, the human perception of risk often influences how risk is addressed [3]. For example, on the level of decision makers the risk perception does not necessarily map directly to probability and severity in a linear fashion. That makes computation of total risk additionally difficult and subjective.

For all these reasons, the concept of a Composite Risk Index (CRI) that could measure the performance of European ATM systems as a whole or also its individual entities (service providers or Member States) is proposed. In simple words, CRI presents a cumulative risk value calculated aggregating all reported, assessed and severity classified safety-related incidents to form an index. This measure of risk exposure is based on probability and severity that considers the human perception of equivalent risk.

1.1 Data

In order to calculate composite risk, each historical, reported occurrence had to have assigned severity and probability. Safety information about reported events was acquired through EUROCONTROL Annual Summary Template (AST) reporting system [4]. The AST reporting mechanism captures information on Air Traffic Management (ATM) related occurrences, both ATM operational and technical occurrences. The safety data, related to the reported occurrences in the AST, included occurrence category (accident or incident) and its severity reported by the States and calculated using severity classification risk assessment methodology (RAT).

For definition of the Weights, that would explain the human perception of risk for each type of safety occurrence, and also overall CRI methodology development in generall, the AST data was provided by EUROCONTROL AST Team for period 2015-2018. Nevertheless, it has to be noted that modelling of Weights can be customised additionally to a local environment, it can be performed using different source of safety occurrences data, as long as the input satisfies the minimum data requirements (This includes, the total number of occurences per each type of incident, separately for each severity class that is modelled. In addition exposure data in terms of flight hours is needed as well.).

The classification scheme for safety occurrences in ATM specifies six severity categories for ATM related occurrences impacting the safe operations of the aircraft [5]. They are as follows: Accident, Serious Incident (AA / A), Major incident (B), Significant incident (C), Not determined (D), No safety effect (E).

The RAT classification scheme [6] specifies five qualitative frequency categories (repeatability), however, these values are not commonly reported through the AST. Moreover, each State in principle should develop their own quantitative boundaries, which should consider national traffic volumes and specific operating conditions of the national ATM system. As these values were not available the occurrence probability was calculated using historical data (frequency of occurrences over all available years was used as a proxy for probability) from the past three years separately for each State in order to simulate/take into consideration local conditions.

1.2 Methodology

As a proxy of safety risk within certain airspace or a State, at the preliminary stage of development, it was decided to base CRI calculations on the following (Figure 1):

- Accidents,

- Operation occurrences - OPS (high/medium risk incidents with Severity A to C): runway incursions (RI), separation minima infringements (SMI), unauthorized penetrations of airspace (UPA),
- Other operational occurrences - OTHER;
- Technical occurrences – TECH (high/medium risk incidents with Severity AA/A to C).

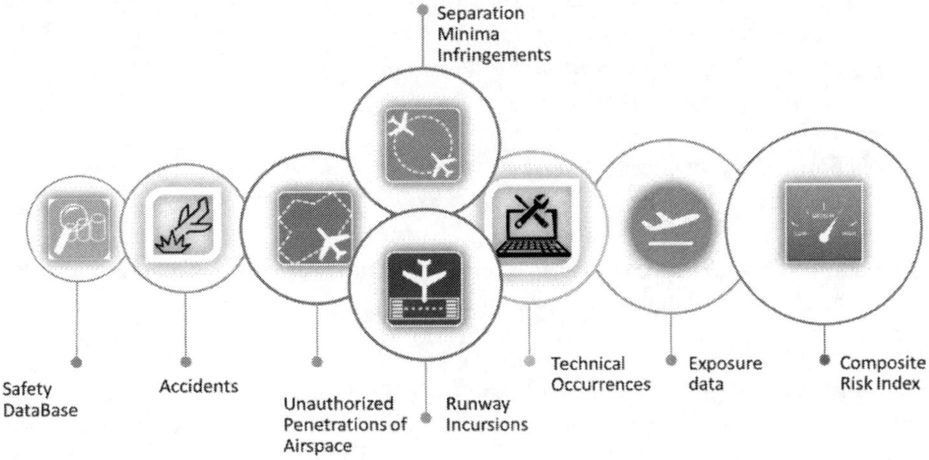

Figure 1. Iconographic showing different elements of data input for calculation of CRI

To take into account the local conditions within each Entity, to have an objective comparison across small and large States/entities, scaling of variables by an appropriate size measure, in this case the total number of flight hours within each State, was used as an additional input (CRI normalised results).

1.2.1 Data input for calculation of CRI

Missing data often hinder the development of robust composite indicators. Data can be missing in a random or non-random fashion. In case of AST safety data available, considering the type of safety information collected, the missing values do not depend on the variable of interest or on any other observed variable in the data set. In other words, the missing values in Severity classification would be of the missing completely at random type: i.e. Severity classification has no correlation with type of occurrence or with reporting entity.

There are two general methods for dealing with missing data: case deletion, or imputation. No imputation model is free of assumptions and the imputation results should be thoroughly checked for their statistical properties, such as distributional characteristics, as well as heuristically for their meaningfulness[2].

Data imputation could lead to the minimisation of bias and the use of 'expensive to collect' data that would otherwise be discarded by case deletion. The uncertainty in the imputed data should be reflected by variance estimates. This makes it possible to take into account the effects of imputation in the course of the analysis. The multiple imputation method, which provides several values for each missing value, can more effectively represent the uncertainty due to imputation [7].

For all these reasons, all severity unclassified/not assessed events (Severity Category D) were distributed into groups A to E based on historical distribution (determined using the last three years of AST data). The probability of occurrence being assigned to specific severity category was calculated using historical data, separately for each State in order to simulate/take into consideration local conditions.

1.2.2 Estimated numbers of occurrences

The formula how the numbers of occurrences of specific type were estimated is presented below.

$$No\_occ\_est_i = \left((No\_D_i * P_i) + No\_occ_i\right) \quad (1)$$

$No\_occ\_est_i$ – *estimated number of safety occurrence type i*
No_D_i – *number of reported occurrences of Severity Classification D (not determined) for occurrence type i*
P_i – *probability of safety occurrence type i*
No_occ_i - *number of reported safety occurrences of type i*

Probability (probability is taken as proxy for frequency as explained before) of each type of occurrence was calculated using the simple principle:

$$P_i = \frac{No_{occi}}{Total\ No_{occi}} \quad (2)$$

No_{occi}– *number of reported safety occurrences type i in a year*
$Total\ No_{occi}$ - *total number of reported safety occurrences type i in a group in all years*

1.2.3 Human perception of risk

In order to add human perception of risk to the CRI index, certain values/weights had to be utilised in order to attribute personal perception of risk to its values.

Overall, when used in a benchmarking framework, weights can have a significant effect on the overall composite indicator and the entity rankings. A number of weighting techniques exist however, regardless of which method is used, weights are essentially value judgements. While some analysts might choose weights based only on statistical methods, others might reward (or punish) components that are deemed more (or less) influential, depending on expert opinion, to better reflect policy priorities or theoretical factors. In case of CRI both statistical/optimisation technique and expert judgement are used.

Accepted methods of quantifying severity include monetary amounts. However, although expressing severity in terms of cost establishes consistency, it is still difficult to put an amount on human life or injuries, or failure or loss of certain functionalities of the system. Furthermore, perception of what constitutes "high" risk may vary from entity to entity and State to State.

Therefore, introduction of Weights to express severity of event allows their description in non-monetary terms which have meaningful and easy understandable explanation in human perception.

Each weight value for specific Severity category was determined using optimisation technique, with the aim to select combination of weights that will not disturb the computation of the CRI from year to year if significant changes in reporting are introduced. In other words, the goal was to determine which combination of weights would result in the lowest standard deviation of CRI values between the years for each State.

Due to a large number of variables involved and enormous number of combination possible, optimisation and selection of Weights was done in several stages:

1. Selection of Weights for accident, all OPS and TECH occurrences (Base Weights);
2. Selection of Weights for OPS occurrences based on their type (RI, SMI, UPA, OTHER), taking into consideration overall OPS Weight determined in Step 1 (Occurrence type Weights);

3. Selection of Weights for OPS and TECH occurrences based on their severity (AA/A, B, C), taking into consideration overall OPS Weight determined in Step 2 (Occurrence severity Weights).

Overall optimal solution in each step, was the one that results in the minimum mean value of CRI standard variations for all entities in a single configuration (combination) of Weights.

In addition, each type of weight selection had predefined weight ranges (based on EUROCONTROL experts judgement, using techiques such as brainstorming and voting) to allow for incremental Severity classification order based on human perception of risk (from accident to Severity C incident, i.e. from high risk to low risk). In other words, each range had an expectation value associated with it.

The following ranges used for selection of Weights in different steps are presented below:

- Base Weights selection:
  - Accidents weight: w_acc = (0.5 : 0.7)
  - Operational occurrences weight: w_ops = (0.3 : 0.5)
  - Technical occurrences weight: w_tech = (0.05 : 0.4)
- Occurrence type Weights selection:
  - UPA weight: w_ops_upa = (0.15 * w_ops : 0.25 * w_ops)
  - SMI weight: w_ops_smi = (0.4 * w_ops : 0.55 * w_ops)
  - RI weight: w_ops_ri = (0.25 * w_ops : 0.35 * w_ops)
  - OTHER operational weight: w_ops_other = (0.05 * w_ops : 0.1 * w_ops)
- Occurrence severity Weights selection (where the letters in the name represent occurrence type and Severity class respectively):
  - w_ops_ri_a = (0.3 * w_ops_ri : 0.7 * w_ops_ri)
  - w_ops_ri_b = (0.1 * w_ops_ri : 0.5 * w_ops_ri)
  - w_ops_ri_c = (0.05 * w_ops_ri : 0.3 * w_ops_ri)
  - w_ops_upa_a = (0.3 * w_ops_upa : 0.7 * w_ops_upa)
  - w_ops_upa_b = (0.1 * w_ops_upa : 0.5 * w_ops_upa)
  - w_ops_upa_c = (0.05 * w_ops_upa : 0.3 * w_ops_upa)
  - w_ops_smi_a = (0.3 * w_ops_smi : 0.7 * w_ops_smi)
  - w_ops_smi_b = (0.1 * w_ops_smi : 0.5 * w_ops_smi)
  - w_ops_smi_c = (0.05 * w_ops_smi : 0.3 * w_ops_smi)
  - w_tech_aa = (0.4 * w_tech : 0.5 * w_tech)
  - w_tech_a = (0.3 * w_tech : 0.4 * w_tech)
  - w_tech_b = (0.1 * w_tech : 0.3 * w_tech)
  - w_tech_c = (0.04 * w_tech : 0.08 * w_tech)

Optimisation results indicated that, for this setup, the best combination of Weights was as in Table 1.

Using estimated number of occurrences and adding human perception of their risk it was possible to calculate CRI for each EUROCONTROL Member State separately. The simple formula to calculate CRI is presented below.

$$CRI_i = No_{occ\_est\_i} * w_i \tag{3}$$

$$CRI = \sum_i \frac{CRI_i}{Total\ N_{occ}} \tag{4}$$

CRI_i – CRI index for occurrence type i
W_i – weight (based on severity and human perception) assigned to specific type of safety occurrence
$Total\ N_{occ}$ – total number of all occurrences in a year

In the formula above *Weights* added to each equation represent additional human perception of risk for specific event, introduced so that the CRI can at the end consider the human perception of equivalent risk.

Finally, to allow applicability of CRI to airspaces with different traffic levels, the CRI was normalised by flight hours for each State. The CRInorm was calculated based on the following formula:

$$CRInorm = \frac{CRI}{Number\ of\ Flight\ Hours} \quad (5)$$

Type of occurrence	Weight index	Value
Weight for accident	w_acc	0.54646
Weight for OPS incident	w_ops	0.38844
Weight for TECH incident	w_tech	0.06509
Weight for RI incident	w_ops_ri	0.09711
Weight for SMI incident	w_ops_smi	0.21364
Weight for UPA incident	w_ops_upa	0.05827
Weight for OTHER OPS incident	w_ops_other	0.01942
Weight for serious RI incident (A)	w_ops_ri_a	0.06798
Weight for major RI incident (B)	w_ops_ri_b	0.02428
Weight for significant RI incident (C)	w_ops_ri_c	0.00486
Weight for serious SMI incident (A)	w_ops_smi_a	0.11196
Weight for major SMI incident (B)	w_ops_smi_b	0.09100
Weight for significant SMI incident (C)	w_ops_smi_c	0.01068
Weight for serious UPA incident (A)	w_ops_upa_a	0.03046
Weight for major UPA incident (B)	w_ops_upa_b	0.02453
Weight for significant UPA incident (C)	w_ops_upa_c	0.00328
Weight for serious TECH incident (AA)	w_tech_aa	0.02980
Weight for serious TECH incident (A)	w_tech_a	0.02597
Weight for major TECH incident (B)	w_tech_b	0.00651
Weight for significant TECH incident (C)	w_tech_c	0.00281

Table 1. Weights based on severity and human perception

2 Preliminary Results

Using methodology described in previous section, the CRI for all EUROCONTROL Member States for 2018 (for which data was available, 39 Member States) is calculated and shown at Figure 2. It shows that two thirds of the EUROCONTROL Member States in 2018 had higher risk exposure in comparison to the EUROCONTROL average. Nevertheless, this calculation did not take into account the local conditions. Therefore, CRI was normalised, (CRInorm), taking into account the total number of flight hours within each State (Figure 3.), as it is assumed that amount of traffic could impact level of risk exposure within specific airspace.

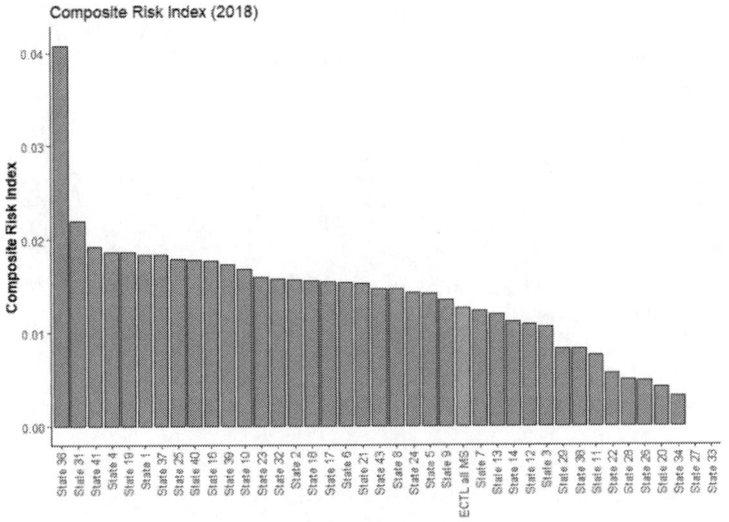

Figure 2. CRI in EUROCONTROL Member States (2018)

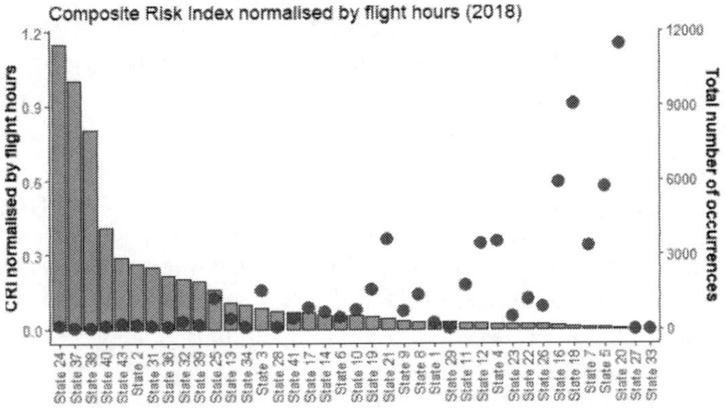

Figure 3. CRInorm vs Total Number of Occurrences in EUROCONTROL Member States (2018)

The further analysis of 2018 CRInorm results indicate that 25% (10 States) of EUROCONTROL Member States have CRInorm above 0.18 (treshhold taken from boxplot of CRInorm, upper quartile of 75% of population). With only three States having high CRInorm (boxplot outliers – i.e. above 0.5). One possible reason for this overall positive result could be that this is somehow related to the reporting culture of the State. Therefore, CRI normalised was correlated with the total number of reports reported by each State. Figure 3. shows that the States with a higher number of reports, which could indicate a good reporting culture (red dots - total number of reports), tend to have a low CRI normalised (blue bars).

2.1 CRI trend

Using CRI index, it is possible to follow the trend of safety performance, as CRI can be used as a quick indicator of the status of either safety performance based on the type and severity of historical reported occurrences but also as indicator of reporting culture. Figure 4. shows how the CRInorm has changed in a period 2015-2018 in EUROCONTROL area.

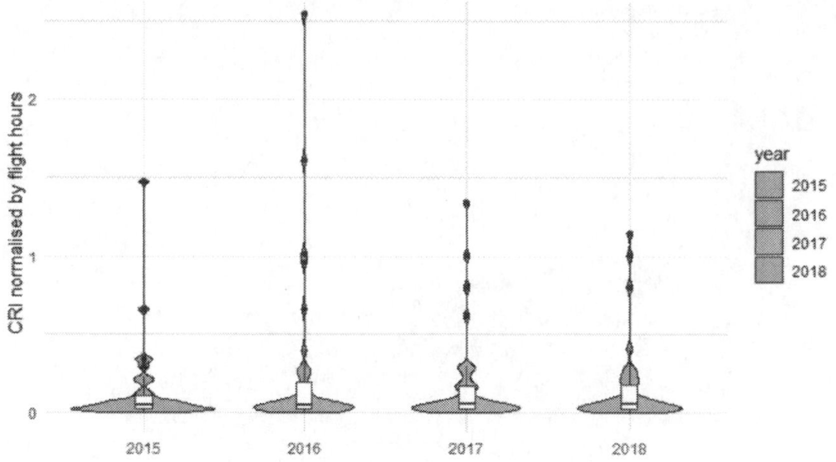

Figure 4. Normalised CRI in EUROCONTROL area (2015-2018)

In 2016 CRInorm has increased comparing to the previous year, which was mainly influenced by very high CRInorm of several States. In 2017, regardless of the massive increase in the number of OPS occurrences by 65% (which have higher Weight values) and overall increase in all reported occurrences by approximately 38%, the CRI scores for EUROCONTROL area have decreased. This trend continued in 2018 as well. Furthermore, on the positive note, the number of States with extreme CRInorm has decreased (number of outliers has decresed).

Overall, using selected weights, further analysis showed that variation in reported number of occurrences does not have adverse effect on CRI computation and does not solely depend on the number of reported events. This is very important as the nature of CRI computation also allows calculation and monitoring of CRI of a single specific type of occurrences, e.g. the key risk occurrences within an airspace or organisation.

3 Conclusion

Idea behind CRI is that the performance of safety system can be analysed within three important broad categories: the quality of reporting system within reporting entity, measured risks within the system, and human perception of risk. Therefore it was considered that the concept of a CRI, as a cumulative risk value calculated aggregating all reported, assessed and severity classified safety-related incidents, has potential to become a proxy of exposure to risk within certain airspace for top management information and decision making.

Preliminary analysis shows that CRI has an ability to allow reporting on the safety performance of the whole European ATM system, but also on the level of its individual entities, e.g. Member States or even at the level of service providers. Moreover, scaling possibility allows measurement of CRI of individual types of safety occurrences as well.

The CRI however, should not be construed as an absolute measuring stick. It is only as good as the fidelity of the data that supports it. In general, specific probabilities of occurrence are not precisely known, and there is some subjectivity in the assessment of severity of the occurrence.

As mentioned before, besides the fact that CRI methodology can be customised to local environment, i.e. Weights can be re-modelled using local safety data, CRI methodology can be scaled up or down to satisfy monitoring of individual entities.

Based on individual local safety data availability, the CRI calculation can be improved by using higher granularity of safety-related data used to compute CRI. In other words, by using safety data with higher granularity, so that Weights are computed separately for each different type of occurrence, (e.g. providing separate weights for different OTHER types of OPS occurrences).

Moreover, initial ranges of different Weights could be fine-tuned based on collective expert opinion. Adjustment of proposed Weights could be further improved via dedicated expert group, both locally and within aviation community. This would also help to better understand potential concept limitations and added value.

Finally, the CRI normalisation could also be done per different metrics, in order to allow inclusion of airspace size, capacity and/or complexity (for example, normalization per sector or number of flights). This could allow adding additional local specific operating conditions into equation.

References

1. Burns, M. (2016). Logistics and Transportation Security, A Strategic, Tactical, and Operational Guide to Resilience. CRC Press, Taylor and Francis Group.

2. Nardo, M., Saisana, M., Saltelli, A., Tarantola, S., (2005). "Handbook on Constructing Composite Indicators: Methodology and User Guide". OECD Statistics Working Papers, STD/DOC (2005). Organisation for Economic Co-operation and Development.

3. Slovic, P. "Perception of Risk". (2016). Available via: https://www.researchgate.net/publication/325954197_The_perception_of_risk

4. EUROCONTROL, (2009). "Reporting and Assessment of Safety Occurrences in ATM," EUROCONTROL, Brussels.

5. EUROCONTROL, (1999). "Severity Classification Scheme for Safety Occurrences in ATM," EUROCONTROL, Brussels.

6. EUROCONTROL, (2013). "RISK ANALYSIS TOOL Guidance Material," EUROCONTROL, Brussels.

7. Jakobsen, J.C., Gluud, C., Wettersley, J., and Winkel, P. (2017). When and how should multiple imputation be used for handling missing data in randomised clinical trials – a practical guide with flowcharts. BMC Medical Research Methodology Volume 17, Article number: 162

Knowing versus understanding: adjusting the contextual lens in safety science

Janice R. McCall[1,*]
[1]Saint Louis University, Parks College of Engineering, Aviation and Technology, United States of America

Abstract. Preparing for the unknown safety threat requires new ways of studying safety in today's highly complex and dynamic environment. Qualitative inquiry offers advantages to expand the contextual lens in safety science by broadening understanding of the environment and conditions people negotiate in normal day-to-day operations. Through the narrative accounts of those with first-hand experience qualitative researchers seek to narrow the gap between work-as-done and work-as-imagined. It is through their individual experience, when analyzed collectively, that a greater context emerges. To aid in determining which research method is best for investigating safety phenomenon, this paper offers an overview of the philosophical underpinnings of qualitative research as well as recent examples from aviation, maritime, and mining.

1 Introduction

In the study of safety, knowing what happened is different than understanding how and why an event occurred. It is here, in the space between knowing and understanding, that interpretive qualitative research contributes a great deal to safety science. Safety science "refers both to what we know about safety and to the ways we have built and continue to build this knowledge" [1]. While qualitative, quantitative, and critical realism all enhance safety research, qualitative research's diversity of ontology, epistemology, and methodology, is well situated to expand context, foster understanding, and strengthening organizational safety learning and resilience. In the complex and dynamic environment, where human life hangs in the balance and safety precariously exists, "sometimes your greatest strength can emerge as a weakness if the context changes" [2].

Of the three research methodologies common in safety science, quantitative, qualitative, and more recently critical realism; quantitative research has dominated. The quantitative approach "has been, and continues to be, the preferred research methodology" [3]. This preference is often driven by the relationship between quantitative research and business management. The management-measurement arrangement capitalizes on the strengths of quantitative research allowing for relatively fast, precise, unbiased, statistical analysis that can determine trends and be generalized to a larger population.

For comparison, a recent quantitative study of seven European airlines, by Kaspers, Karanikas, Piric, van Aalst, de Boer, & Roelen [4] concluded "that there is limited empirical

*Corresponding author: jan.mccall@slu.edu

evidence about the relationship between Safety Management System (SMS) processes and safety outcomes." Yet, these finding lack the detail from accounts of those working with SMS. Similar findings in a Størkersen, Antonsen, and Kongsvik, [5] qualitative study of Norwegian maritime passenger transportation reported that while personal injury had decreased, "ship accidents had increased in the period during which safety management has been regulated." Interviews conducted with 47 regulators, managers, and crewmembers about their experience with SMS revealed negative consequences to navigation such as "unthinking compliance" and that the "administrative burden is easily associated with less concentration on navigation" [5]. Qualitative inquiry captures a glimpse of the "stocks of images in our heads, put there by our origins, upbringing and experience of the world, which we use to make sense of the world" [6].

These two studies reveal a weakness in quantitative research, that is a strength of qualitative. The deductive approach "fails to provide insight into the participants' individual or personal experiences. They do not let the respondents describe their feelings, thoughts, frames of reference, and experiences with their own words" [7]. Qualitative research weaknesses include being time consuming, biased, and seldom generalizable.

These strengths and weaknesses led to the early association of quantitative as hard, fixed, and objective while qualitative was considered soft, flexible, and subjective [8]. Hong and Lang [9] apply the Chinese philosophy Yin-Yang to these two methodologies, where power "comes from capturing the very nature of the world and the mechanism through which the world operates." This is where the overreliance on quantitative research limits our ability to capture the human experience, the messy details, that create the context surrounding safety.

1.1 Purpose

Rather than exam all three methodologies, quantitative, qualitative, and critical theory, this paper attempts to engage the reader in a deeper understanding of the benefits and applications of qualitative research in safety. Safety practitioners, mangers, and anyone interested in the human experience may find this paper worthwhile. It is divided into sections addressing how qualitative research uses individually constructed experience, collectively analyzed, to illustrate context in order to broaden understanding of safety issues. The sections include, the meaning of context as used in qualitative research, the role context plays in knowing and understanding, and a brief review of the philosophical underpinnings of interpretive qualitative. The final section offers advantages of qualitative research and highlights its usefulness in three recent safety studies.

2 Context

"*If an intervention is a seed then the context is the soil*" [10].

Context is a complicated, commonly used term, that carries a wide range of meanings. Rather than delve into the multitude of definitions of context, including 150 analyzed by Bazire and Brézillon [11], covering "various disciplines such as computer science, philosophy, economy, business, HCI [Human Computer Interface]etc." this paper focuses on context as derived from lived experience through qualitative research in safety. Van Dijk [12] argued that "the notion of context as it is used in the social sciences is not a strictly theoretical concept, but rather a more or less fuzzy notion denoting a situational, historical, geographical, social or cultural environment of a phenomenon being studied." Context, in this regard, aligns with social sciences, sociology, and linguistics, by focusing on gaining meaning and a broader understanding of events, including organizational safety conditions and culture.

Safety cannot be understood and explained at a distance from the cultural and social environment in which it appears. To the contrary, safety, like any organizational endeavor, requires a balance between practice (work-as-done), and process (work-as-imagined). Closing the space between the two entails what Brown and Duguid [13] refer to as "collective wisdom," putting individual stories together and drawing from the experience of a group.

Described through qualitative study of lived experiences, we see context returned to its early Latin origins *contextus* meaning "a joining together," or *contexere* "to weave together" [14] where multiple narratives join to create context in safety. The qualitative researcher is not looking for the explanation of a safety event, or the root cause, rather an interpretation of how each person involved experienced the event, and a collective context from which to garner understanding of the organizational safety conditions, culture, and systems in a holistic manner [15].

Each description of a safety event is shaped from the experience within the sociotechnical environment, then interpreted, and meaning constructed, within the minds of the individuals. Technology cannot replace experience, values, and judgement [16]. It is the lived experience and judgment found in capturing and analyzing that yields a broader context. "We collect safety stories because stories provide context. Stories we hear are what storytellers choose to tell because it is important to them. We often learn about safety issues that management or expert consultants had no clue about" [17]. To borrow from Duranti and Goodwin's [18] study of linguistics, what the qualitative researcher "seeks to describe is not what they consider context, for example their map of the city in which the blind man finds himself, but rather how the subject himself attends to and organizes his perceptions of the events and situations he is navigating through."

Such views diverge from traditional positivist interpretations of context, but when used concurrently, offer new insights into safety through organizational systems functions and the environment employees negotiate in day-to-day operations to ensure safety. In the "positivist, rational–analytic philosophy that has always dominated in science and medicine, context is usually defined as an 'objective' phenomenon, something 'real', something tangible and 'out there' – factors, variables, objects, events, domains and so on" [19]. There is risk in such belief that fails to see the human construct of context. Hummerdal [20] characterized this risk when he said; "our belief in best practice, standardized methods, and other versions of the one best way, can be seductive and create a blindness to the contextual adaptations that are required to make things work in everyday performance."

While the definition of context may end where it started, as a complicated, complex, and commonly-used term, its critical importance to advancing safety is clear; success in capturing context in safety is not found in the division of methods, but more accurately, when combined to provide greater understanding. Dixon-Woods [21] proposed focusing on "practical wisdom" and a blend of methods as the best way forward.

2.1 Context in Knowledge and Understanding

Before turning to philosophy, let us first explore how context, drawn from qualitative inquiry, contributes to knowledge and translates to understanding in safety science. Context provides familiarity and connects learning to meaning as part of knowledge development [22]. Safety is dependent on both individual and organizational learning [23-24]. "Cognitive theories tell us that learning occurs not by recording information but by interpreting it" [25]. This type of learning in safety involves gaining 'meaning' from knowledge. Safety learning, or any

learning, is "knowledge-dependent; people use current knowledge to construct new knowledge" [25].
Knowledge, like context, is not easily defined. Nonetheless, in safety it may be useful to view knowledge as "an ordering and organization of a world constituted by our experience" [26]. Piaget's [27] theory of constructivism argued that "to know an object, to know an event, is not simply to look at it and make a mental copy or image of it." Piaget considered knowledge as acquired through cognitive schema that formed structures, categories, and transformed objects based on experience. Perhaps the easiest way to think of knowledge is to apply Perkins' [28] definition; "Knowledge is information on tap."

To expand insight into the safety environment, qualitative inquiry must consider the person(s) knowledge of events, what it meant individually and collectively, and how they understood the event to shape the collective context. Figure 1 portrays how individual context is translated into meaning, knowledge, and understanding, that when gathered and analyzed through qualitative inquiry presents a broader context for study. Each circle represents the individual's development of understanding through context, meaning, and knowledge within the web of safety culture. While each experience is valuable, it is the combined analysis that provides collective context for organizational safety learning and understanding.

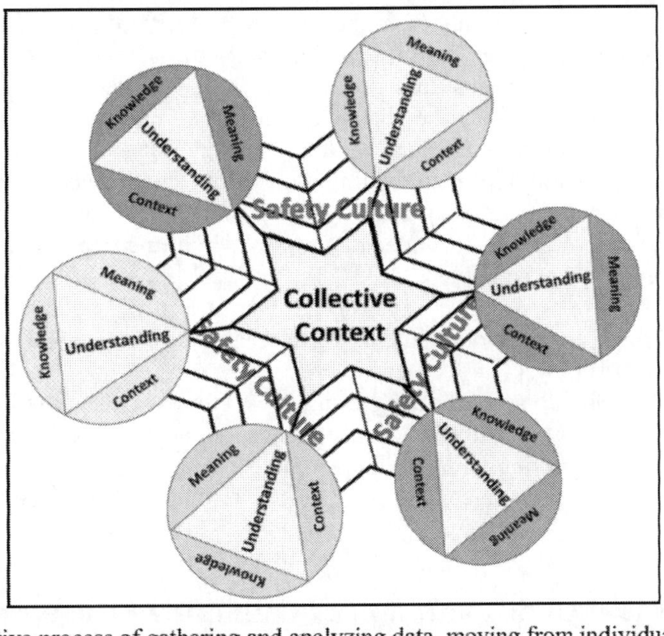

Figure. 1. Qualitative process of gathering and analyzing data, moving from individual context to meaning, knowledge, understanding, and collective context.

Understanding takes many forms in the qualitative research process. It begins with collection of individual data from the first-hand experience of those involved, in the form of interviews, observations, questionnaires, and narratives accounts. The analysis process then builds to a collective understanding through review of all those involved. During the analysis process, themes emerge that redefine and clarify context for further safety studies. Qualitative analysis is like putting together a jigsaw puzzle without the picture on the box. The qualitative researcher has a general idea there is something to be found within the data, based on the experience of those involved, but cannot interpret it without seeing how the pieces work together to form collective themes. Collective themes expand our understanding and reshape the contextual lens to identify where the gaps exist between work-as-imagined and word-as-done.

Safety today is perhaps more challenging than ever, existing in complex and dynamic environments, that require new approaches. The ability to rely on past performance as an indicator of future success is limited; current safety theorists have turned to capacity, reliability, and resilience as key indicators of safety's future performance [29-32]. Knowing is not enough, we must expand our understanding in order to build flexibility and capacity into safety systems. It is time to revisit understanding as Perkins' defined it in 1998; "Understanding is the ability to think and act flexibly with what one knows." Understanding is a "flexible performance capability" [28] that allows individuals and organizations to adapt and quickly react to unexpected safety events.

In safety it is not always the unknown that presents the greatest threat, it is what is known and not shared. Often, these are the day-to-day adaptations we don't understand, that people make to ensure safety and efficiency, and which lie outside the very detailed and structured scope, Standard Operating Procedures (SOPs), checklists, policies, and regulations. "Work gets done because of people's effective informal understandings, their interpretations, their innovations and improvisations outside those rules" [33], which you will see in the case studies at the end.

This begs the question at the core of the Safety Differently and Safety II movement; If the same employees are doing the same work safely and successful 99.9% of the time, why is our attention focused on learning from the random and unpredictable failure? Hollnagel, Wears, & Braithwaite [34] call for a new definition of safety gets to the heart of the importance of expanding understanding.

The definition of safety should be changed from 'avoiding that something goes wrong' to 'ensuring that everything goes right'. Safety-II is the system's ability to function as required under varying conditions, so that the number of intended and acceptable outcomes (in other words, everyday activities) is as high as possible. The basis for safety and safety management must therefore be an understanding of why things go right, which means an understanding of everyday activities [34].

By listening to workers unique experiences, we enter a perspective in which we see normal humans doing normal work while trying to create success amongst scarcity, imperfect and conflicting setups [35].

The goal of listening to and collecting employees' stories, as Conklin [36] asserts, is "to focus learning on the context before the consequence—the context is the only place you can have an impact on improving your organization." Moving away from safety outcomes as resultant and toward safety outcomes as emergent is one of the strengths of qualitative research.

3 Philosophical Underpinnings of Qualitative Research

To some, the philosophical underpinnings of qualitative research may seem unimportant in the study of safety, burdensome, and dated. Nonetheless, there are philosophical differences of critical importance to safety. Highlighted below are certain distinctions in research paradigms along with selected preeminent philosophical foundations guiding safety science.

The situated researcher approaches the world with a set of ideas, a framework (theory, ontology) that specifies a set of questions (epistemology) that are then examined (methodology, analysis) in specific ways Denzin & Lincoln [37]. Understanding this difference when determining which approach would be most useful, either by deductively testing safety theories or inductively searching for new theory. In Figure 2, Kyrö [38] depicts how understanding paradigm as a mediator, a gathering link, between philosophical bases and methods, might help in choosing methods and theories for study.

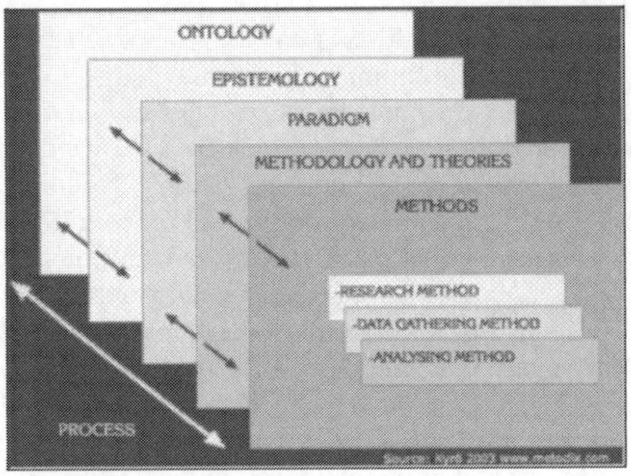

Figure. 2. Hierarchical approach to paradigms [38].

3.1 Context in Knowledge and Understanding

Embracing the variety of research paradigms moves beyond the linear trappings of root cause analysis and toward understanding the increasing complexity of modern safety environments. Paradigms are "basic belief system based on ontological, epistemological, and methodological assumptions" [39].

- Ontology: Assumptions about the nature of reality
- Epistemology: How the researcher comes to know that reality
- Methodology: How the researcher accesses and reports what is learned about reality

The directional relationship between these philosophical assumptions can be seen in the work of [40], Figure 3.

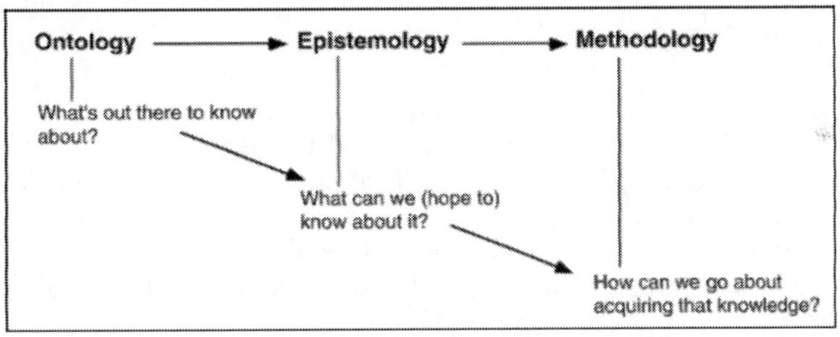

Figure. 3. Ontology, epistemology, methodology; directional dependence [40].

A recent web search revealed as many as 16 paradigms currently in use, although two remain dominant in safety research, Positivist, Interpretivist/Constructivist [41-46]. There is also a growing community of researchers, under the critical realist paradigm, that argue "scientific knowledge does not consist of 'propositions' or 'statements' about events or phenomena; rather, it is comprised of (provisional and fallible) descriptions of structures and powers" [47]. A brief review of the origins of research paradigms follows.

The word "paradigm" stems from late 15th century, Late Latin *paradigma* and Greek *paradeigma* meaning a "pattern, model; precedent, example" [14]. In scientific communities, American physicist, historian, and philosopher, Thomas Kuhn's [48] critical work *The Structure of Scientific Revolutions*, described paradigms as "universally recognized scientific achievements that for a time provide model problems and solutions to a community of practitioners." Yet, it was the publication of Ludwik Fleck's [49] *Genesis and Development of a Scientific Fact*, that introduced "thought collectives" which enlighten Kuhn's definition of paradigms and work on scientific communities of knowledge. "When we look at the formal aspect of scientific activities, we cannot fail to recognize their social structure" [49]. The social scientists' contribution to safety science must be acknowledged as it has increased our understanding of human, organizational, and socio-technical factors influencing safety.

Building on the work of Kuhn and Fleck, Guba and Lincoln [50] define paradigm as "the basic belief system or worldview that guides the investigator, not only in choice of method but in ontologically and epistemologically fundamental ways." When considering research in safety science the paradigm encapsulates the underlying logic and theoretical assumptions that support selection of study methodology. Without a paradigm. Kuhn [48] warned, that the "reason for seeking some form of more recondite information, early fact-gathering is usually restricted to the wealth of data that lie ready to hand."

Data is seldom readily available in qualitative studies, nor easily collected, because the researchers seek understanding in the depth and richness of human experience. Understanding the world, and humans in it, leads to the next section an overview of preeminent philosophers associated with the foundational ontological and epistemological differences in safety science.

3.2 Preeminent Philosophers

3.2.1 The Ontological Divide

The ontological divide originates in Plato's Protagoras (490-420 B.C.) where Socrates engages in discussion with one of the Sophists, Protagoras. Plato establishes the relativist position through two famous quotes by reference to Protagoras' book:

- "One can only think what he experiences, and what he experiences is true." [51]
- "Of all things the measure is Man, of things that are, that they are, of things that are not, that they are not" [52].

Plato (427 – 347 B.C.), drawing on his experience as a student of Socrates (469-399 B.C.), and the work of Heraclitus (530-470 B.C.) and Parmenides (515–450 B.C.), developed a different philosophy of reality. Plato's view, in contrast with Socrates, was that humans were fallible, limited in our abilities, our senses were inadequate to interpret reality, which he described as Form, or the perfect idea, existing only in the spiritual world [53]. These ontological positions are seen in safety science today as Interpretivism and Positivism respectively.

3.3 Epistemology divide

Later, Aristotle's *Metaphysics*, Books I-XIV, [54], introduced "the first principle of all things" and sets the course of inquiry into What is a thing? Here Aristotle diverged from Plato's view of reality as Forms, existing outside the physical world. Aristotle, more concerned with practical wisdom [Greek, *phronesis*] argued the world exists in observable physical arrangement "what is actually there… the essence of the thing" but the only way humans experience the world is through our senses, giving humans contact and, at the same time, insulating us from the world [55]. The disagreement between Plato and Aristotle on the meaning of knowledge is relevant still today.

3.3.1 Objectivism-Positivism takes center stage – Galileo

Nearly 2000 years after Aristotle, Galileo (1564-1642), known as The Father of Modern Science, "rejected metaphysics in favor of a secure method of generating knowledge" [56]. In *Siderius Nuncius* [The Starry Messenger] Galileo (1610) made systematic observations of the Moon, stars, and moons of Jupiter. His methods stood in contrast to Aristotle and the Catholic Church's prevailing geocentric ideology Jakobsen [57]. Galileo's view of the world as objective, existing outside of human reality, as an observable existence to be studied, contributed to the growth and dominance of the positivist paradigm. His view was supported by Descartes, Hobbes, and Leibniz emphasizing the world as elements and calculations Palmer [58]. Over time positivism has changed but the belief in one objective truth remains.

3.3.1.1 Postpositivism

Postpostivism, born from the work Popper's [59] theory of falsifiability, employs similar research methods but accepts that bias cannot be fully removed. "No matter how faithfully the scientist adheres to scientific method research, research outcomes are neither totally objective, nor unquestionably certain" [60]. Truth, for the postpositivist is no longer objective but approached with a measure of skepticism where reality exists outside the human experience, but we are limited in our ability to acquire it. Constructivist, which may include postpositivist, find reality to be a subjective construct shaped by social influence within the human mind. What holds in safety is that both objective and subjective description of events are valuable.

3.3.2 Interpretivism-Constructivism

The last of the distinctions of significance in qualitative research in safety is the introduction of constructivism as a paradigm closely related to interpretivism. The two are frequently used interchangeably. For the interpretivist, truth and knowledge are not external to human experience, and "social meaning is not automatically located within an object or situation, it is created by human beings" [61]. As mentioned previously, meaning is the connection between context and creating knowledge, and understanding. It is the creation of meaning that leads to constructivist theory.

Jean Piaget (1896–1980), Swiss biologist and psychologist, along with Lev Vygotsky (1896–1934), Heinz von Foerster (1911–2002), and John Dewey (1859-1952), are recognized as the fathers of constructivism. Their combined contribution has reshaped the epistemological framework that supports continued use of qualitative research and its expansion into safety science. Constructivist theory "defines knowledge as temporary, developmental, socially and culturally mediated, and thus, non-objective" [62]. Piaget posited that "humans cannot be given information, which they immediately understand and use; instead humans must construct their own knowledge" [27]. The interpretivist/constructivist view of the world accepts multiple realities and therefore understands how many people experiencing the same event may all see and remember it differently. Rather than being a detractor, this broadens the context of safety events allowing for further exploration and greater understanding.

3.3.3 Paradigms are not permanent

Paradigms provide a "general philosophical orientation to research and can be combined or used individually" [41]. The type of research questions developed in safety studies lead the researcher toward a specific theoretical approach. In the qualitative heuristic approach, Abbott [63] argued "the idea of heuristics is to open new topics, to find new things. To do that, sometimes we need to invoke constructivism. Sometimes we need a little realism." Additionally, paradigms are constructed through social behavior, therefore "fluid and how we think or behave cannot be completely compartmentalized with clear-cut boundaries. As such, to think about a paradigm as a worldview or epistemological stance does not preclude the cross-over of ideas" [44]. The ability to adapt and adjust paradigms is one of the benefits of qualitative inquiry allowing researchers to explore phenomenon from many angles.

4 Qualitative Inquiry Benefits and Case Examples

4.1 Benefits

There are several benefits associated with qualitative research in safety science, key among them are, understanding individual and organizational behavior, predictive and emergent theories, flexibility, and acceptance of multiple truths. There are also some disadvantages, such as being highly subjective, finding is not easily graphed and should be tailored for decision makers, and data is not intended to be replicated. Nonetheless, qualitative inquiry is an excellent tool used independently or in conjunction with quantitative, critical realism, or mixed method studies.

Helman [64] reported that qualitative methodologies "provides the most useful means of understanding why people behave in certain ways and for exploring the relationship between beliefs and behavior." Unlike other methods, qualitative research, using open-ended questions, is uniquely positioned to gather narrative data providing deep and rich information about the environment and conditions surrounding safety work. Hummerdal [35] addresses the importance of capturing stories that "draw attention to things that are entangled and emergent." His use of the term "messy details" describes the complex safety environment in modern socio-technical systems. "It brings out the messy details and other real aspects that impact how work is carried out. Relationships, feelings, mess, smells, perspiration, sounds, screaming, anxiety, pain, joy, pride and camaraderie are all highly unlikely to be captured by big data. Yet, all these aspects impact work, and safety" [35]. Qualitative research also reveals critical safety information or what Love, Smith, Ackermann, & Irani [65] refer to as "uncomfortable knowledge and inconvenient truths that can confront projects in everyday practice'"

Another key feature of qualitative research is the focus on developing predictive or emerging theory. Charmaz [66] notes that "qualitative researchers have one great advantage over our quantitative colleagues. We can add new pieces to the research puzzle or conjure entire new puzzles— while we gather data —and that can even occur late in the analysis. The flexibility of qualitative research permits you to follow leads that emerge." While frequently associated with the grounded theory method, emergence may be found in any interpretive study, especially those employing open-ended questions. "By creating open-ended, non-judgmental questions, you encourage unanticipated statements and stories to emerge" [66] that often reveal latent or unknown safety threats.

Through qualitative data collection and analysis multiple truths may emerge. "Researchers conducting qualitative studies embrace the ontological assumption of multiple truths or multiple realities, i.e., that each person has an understanding of reality from an individual perspective" [67]. Each individual shapes memory and meaning independently, no two experiences are identical. The qualitative researcher's role is not to ascertain which account is the more accurate or 'truthful,' or who is more 'believable.' Instead, to understand how 'situated accounts' are told in a way that allows the speaker to achieve a different purpose through emphasizing some aspect of their stories and deemphasizing others [68].

4.2 Examples of Qualitative Benefits in Completed Studies

Three recent examples are included to provide the reader with insight into the benefits of qualitative inquiry in safety. These examples demonstratives one or more of the benefits of qualitative research. Note the flexibility, adaptability, and emergent theories within these safety studies.

4.2.1 Finnish Maritime

A Finnish maritime safety, conducted by Teperi, Lappalainen, Puro, and Perttula [69] offers a good example of the emergent property of qualitative research. Amongst the many findings revealed from analysis of data collected from "17 operative employees and 12 safety and unit managers" was that inspections conducted by maritime authorities focus almost exclusively on technical matters with little attention paid to psychosocial factors and Human Factors. These types of emergent factors are valuable to organizations, and regulators, in developing effective safety policies and practices, including inspection reports.

This study revealed that "maritime safety culture has not yet proceeded from Safety-I thinking to Safety-II." During the study, the authors explained that several participants defined safety reporting as weak. One of the participants is quoted saying "that (reporting incidents) has been quite weak. If something really serious happens, only then we report, but otherwise it's quite sparse." The maritime study deftly brings to the surface participants individual experience, their situated understanding of events, and contributes a great deal to the larger picture of safety. This type of context is essential to developing organizational or industry-wide safety intelligence.

4.2.2 Australian Mining

Equally important, and a strength of qualitative research, is when successful measures and practices emerge from the research. Such was the case in Havinga, Dekker, and Rae's [70] study of blast crews within an Australian mining operation. One aim of their "study was to investigate the possibilities of crew resource management training for blast crews in mining." To the researchers' surprise, what they found was "practitioners purposely left 'traces' for themselves and other crewmembers in the environment" and "a lot of communication was happening nonverbally, although not necessarily implicit" As reminders, member of the blast crew would leave stones to alert other team members to hazards, along with other forms of non-verbal communication. The actions of the blast crew were not codified in procedures or policy but developed as team norms. In this study flexibility and emergent aspect of qualitative research contributed to broadening context and heightened understanding, which in turn served to enhance organizational safety by increasing adaptive capacity and resilience.

4.2.3 Aviation

Lastly, I offer an example from aviation through Lai, Chen, Khoo, & Zheng [71] qualitative study of unstable approaches to landing. "Executing a successful landing requires meeting all the stable approach criteria, including proper airspeed, altitude, power setting, etc. Otherwise, an unstable approach can occur." They examined shared mental models of unstable approaches using critical decision method to conduct semi-structured interviews of nine airline pilots and seven air traffic controllers. The significance of this qualitative study is apparent considering "the safety data from the IATA GADM Accident database show that the approach and landing phases of flight account for the major proportion of all commercial aircraft accidents; 65% of the total accidents recorded from 2011-2015. Unstable approaches were identified as a factor in 14% of those accidents" [72].

The researchers explored how informational, operational, evaluative, and team-related mental model disconnects, occur between pilots and air traffic controllers that results in inconsistent shared mental models and increased risk of unstable approaches. Six mental model disconnects for pilots and five for air traffic controllers were identified. I cannot cover the paper in detail, and leave that to the reader to pursue, but include one prominent conclusion that highlights how qualitative research increases context in safety studies. "Based on the results obtained, the unequal information sharing is the major factor which causes task-related operational disconnects…Therefore, the improvements in information sharing in the landing system deserve more attention" [71].

Through the narrative accounts of these pilots and air traffic controllers, and with the researchers' openness to multiple truths, the need for increased key information emerged. Pilots with their limited view of the airspace environment seek specific targeted information to shape their understanding, and mental model, to more closely align with the controllers. Together they capture the context of the airspace and aircraft management to ensure safety.

5 Conclusion

Expanding the contextual lens in safety science is about broadening our understanding of the environment and conditions people negotiate in normal day-to-day operations. While all methods of research have a role to play qualitative research is uniquely suited to capture and analyze stories from those with first-hand experience of how work-is-done, and how it differs from work-is-imagined. It is through their individual experience, when analyzed collectively, that a greater context emerges allowing for increased understanding and action.

As revealed in the three recent examples, the researchers' flexibility and willingness to accept multiple truths, coupled with non-judgmental open-ended questions creates a pathway to new knowledge and understanding that builds safety intelligence. There is little question it is time for greater emphasis on using qualitative research, both independently and in conjunction with quantitative studies in today's complex and dynamic safety environment. The time is right to explore new ways of not only looking at safety sensitive work but also listening to those with the immediate experience. Qualitative research in safety science brings the diversity of experience into a collective understanding that supports strong safety culture.

References

1. E. Hollnagel, Safety Sci._**67**, 21 (2014)
2. H. Bhogle. Indian Expr News. (2011).
3. C.V. Ferroff, T.J. Mavin, P. R. Bates, P. S. Murray, *Aeronautica*, **2**, 1 (2012)
4. S. Kaspers, N. Karanikas, S. Piric, R. van Aalst, R.J. de Boer, A. Roelen, PESARO 2017. The 7thIARIA, 9 (2017)
5. K.V. Størkersen, Antonsen, T. Kongsvik, J. Risk Research, **20**, 1154, 1168 (2017)
6. P.B. Checkland, Human Sys. Mgmt, **8**, 279 (1989)
7. K. Yilmaz, K. Eur. J Ed, **48**, 313 (2013)
8. P. Halfpenny, Soc. Review, **27**, 799 (1979)
9. W. Hong, K.R. Lang, *A Philosophical Foundation of Qualitative Modeling Methodologies Based on the Yin-Yang Principle. 134* (2001).
10. J, Carrier, Cardiff Sch. of Health. Sci. 8 (2018)
11. M. Bazire, P. Brézillon, CONTEXT 2005, 31 (2005)
12. T.A. Van Dijk, Discourse and Cont. Soc. Chng. **54**, 285 (2007)
13. J.S. Brown, P. Duguid, *The Social Life of Information: Updated, with a New Preface,*100 (2017)
14. context, paradigm. Etymonline. (2019)

15. T. Greenhalgh, G. Robert, P. Bate, F. Macfarlane, O. Kyriakidou, *Diffusion of innovations in health service organisations: a systematic literature review.* 195 (2008)
16. R. Robinson, Urban Tech: People, Place, Tech. (2015)
17. G. Wong, Safetydifferently.com. (2018)
18. A. Duranti, C. Goodwin, *Rethinking context: Language as an interactive phenomenon* **11**, 4 (1992)
19. P. Bate, *Perspectives on Context.* 8 (2018)
20. D. Hummerdal, Safetydifferently.com. (2018)
21. M. Dixon-Woods, *Perspectives on Context.* 95 (2014)
22. A. Oswalt, M. Dombeck, *Rethinking context: Language as an interactive phenomenon.* **11**, 300 (2005)
23. A.C. Edmondson, Div. of Research, Harvard Bus. 255 (2002)
24. A.K. Cebrowski, J.W. Raymond, *Operationally responsive space: A new defense business model. 77* (2005)
25. L. Resnick, *Knowing, learning, and instruction: Essays in honor of Robert Glaser*. 2 (2016)
26. E. von Glasersfeld, *The invented reality*, 17. (1984)
27. J. Piaget, *To understand is to invent:* The *future of education.* 53 (1973)
28. D. Perkins, The Education Digest. **59**, 39 (1998)
29. T. Conklin, Pre-Accident Podcast. (2019)
30. S. Dekker, *Safety differently: human factors for a new era.* (2014)
31. E. Hollnagel, *Safety-I and safety-II: the past and future of safety management.* (2018)
32. D.D. Woods, E. Hollnagel, *Resilience engineering.* 15 (2017)
33. S. Dekker, HindSight, EUROCONTROL, **25**, 8 (2017)
34. E. Hollnagel, R.L. Wears, J. Braithwaite, *From Safety-I to Safety-II: a white paper.* 23 (2015)
35. D. Hummerdal, Safetydifferently.com (2016)
36. T. Conklin, *Pre-Accident Investigations: Better Questions-An Applied Approach to Operational Learning. 42* (2016)
37. N.K. Denzin, Y.S. Lincoln, *The landscape of qualitative research,* **1**, 23 (2008)
38. P. Kyrö, Metodix. (2002)
39. E.G. Guba, Y.S. Lincoln, *Naturalistic inquiry*. 107 (1985)
40. C. Hay, *Political analysis: a critical introduction*. 64 (2002)
41. J. W. Creswell, J. D. Creswell, *Research design: Qualitative, quantitative, and mixed methods approaches. (2017)*
42. B. Chilisa, *Indigenous research methodologies*. (2011)
43. J. Jericho, The Free School. (2017)
44. C. Kivunja, A. B. Kuyini, Int. J. Higher Ed. **6**, 26 (2017)
45. Y.S. Lincoln, S.A. Lynham, E.G. Guba, *The Sage handbook of qualitative research*, *4*, 97 (2011)
46. D.M. Mertens, *Int. J. of Multiple Research App.*, **4**, 9 (2010).
47. P.S. Gorski, *What is critical realism? And why should you care?* 669 (2013)
48. T.S. Kuhn, *The structure of scientific revolutions*. 15 (1962/2012)
49. L. Fleck, *Genesis and development of a scientific fact.* 42 (1935/2012)
50. E.G. Guba, *Handbook of Qualitative Research.* **11** 105 (1994)
51. G.J. Mattey, UC Davis Philosophy 21. (2001)
52. D.W. Graham, *The texts of early Greek philosophy: the complete fragments and selected testimonies of the major Presocratics*. 703 (2010)
53. L. Slawson, Owlcation. (2018)
54. N. Ross, *Metaphysics.* 73 (2015)

55. B. Gore, *Cosmo Learning*. (2011)
56. J.C. Pitt. *Galileo, human knowledge, and the book of nature: Method replaces metaphysics* 50, 4 (2013)
57. T.G. Jakobsen, Popular Soc. Sci. (2013)
58. D. Palmer. *Looking at philosophy: The unbearable heaviness of philosophy made lighter*. (2009)
59. K. Popper, *The logic of scientific discovery*. (1934/2005)
60. M. Crotty, *The foundations of social research: Meaning and perspective in the research process. (1998)*
61. S. Dyson, B. Brown, *Social theory and applied health research*. 19 (2005)
62. J. Brooks, M. Brooks. *In Search of Understanding: The Case for Constructivist Classrooms*. vii (1993)
63. A. Abbott, *Methods of discovery: Heuristics for the social sciences*. 191 (2004)
64. C.G. Helman, *Primary care research: Traditional and innovative approaches*, 105 (1991)
65. P. Love, J. Smith, F. Ackermann, C. Irani, Int. J. of Project Mgmt. (2019)
66. K. Charmaz, *Constructing grounded theory: A practical guide through qualitative analysis*.14 (2006)
67. C. Erlingsson, P. Brysiewicz, African J. of Emergency Med, **3**, 92 (2013)
68. R. Barbour, *Introducing qualitative research: a student's guide*. (2013)
69. A.M. Teperi, J. Lappalainen, V. Puro, P. Perttula, WMU J of Maritime Affairs, 17 (2018).
70. J. Havinga, S. Dekker, A. Rae, *Theo. Issues in Erg. Sci*, **19**, 213 (2018).
71. H.Y. Lai, C.H. Chen, L.P. Khoo, P. Zheng, Rel. Eng. and Sys. Safety, **185**, 383 (2019).
72. International Air Transport Association (IATA). 2nd, 49 (2016)

MATEC Web of Conferences **314**, 02001 (2020) https://doi.org/10.1051/matecconf/202031402001
ICSC-ISATECH 2019

Metaheuristics optimized machine learning modelling for estimation of exergetic emissions of a propulsion system

Tolga Baklacioglu [1,*], *Onder* Turan [2], and *Hakan* Aydin[3]
[1]Eskisehir Technical University, Faculty of Aeronautics and Astronautics, TR-26470 Eskisehir, Turkey, tbaklacioglu@eskisehir.edu.tr
[2]Eskisehir Technical University, Faculty of Aeronautics and Astronautics, TR-26470 Eskisehir, Turkey, onderturan@eskisehir.edu.tr
[3]TUSAS Engine Industries, Eskisehir, Turkey
e-mail: tei.hakan@gmail.com

Abstract. You This study offers a metaheuristic design for primary parameters and architectures of two models of artificial neural network (ANN) in predicting a business jet aircraft's exergo-emission parameters, such us exergy destruction ratio (rex,dest) and waste exergy ratio (rwex), at different flight stages. In consideration of this, the development of hybrid genetic algorithm (GA)-ANN models has been achieved by considering real databases of rex, dest and rwex at various power levels. Implementing a metaheuristics-based optimization on the generated multilayer perceptron (MLP) ANN models has produced the most favorable initial network weights, step-size, biases as well as training algorithm's back-propagation (BP) momentum rate in addition to optimal quantity of neurons in the hidden layer(s) with regard to the topology design. In accordance with an error assessment approach, there exists a close fit linking the reference real data and rwex (linear correlation ratio, R, value of 0.999851) as well as rex,dest (R value of 0.999985) predicted values.

1 Introduction

With the growth of air travel around the world, environmental impacts and energy consumption of the aviation industry has increasing rapidly. Fuel consumption and aircraft emission are about 3% and 2% of the total fuel usage and total CO2 emissions around the world, respectively. Furthermore, according to the IATA, total greenhouse gas emissions associated with air transportation will be 400–600% higher in 2050 than in four decades [1-4].

Fuel price and energy efficiency of their engine systems effect the total energy efficiency of air transportation systems. For long-term sustainability, many measures are resorted to control wasted emission containing the optimizing fleet and air route, adapting wing-tip fin by aircraft designers and airliners. As energy usage in next generation aircraft fleet gains

* Corresponding author: onder.turan@gmail.com

importance due to operating costs, aircraft must be optimized for energy efficiency, design technologies, environmental considerations, operating and economic costs [5-12].

Environmental problems, i.e. global warming, noise and atmospheric pollution, are related to energy efficiency. Besides, for the sustainable development and environmental sustainability, energy consumption plays a crucial role. So, energy and exergy concepts can be utilized incorporated with environmental considerations, sustainability and thermodynamics. Nowadays there is a great interest in studying the linkages between energy, exergy, environment and sustainability. Exergetic and energetic metrics are useful dimensions for evaluating aircraft environmental and technological performances [13-17].

For improving performance, environmental and economical characteristics of aircraft, powerful engines and less fuel consumption are inevitable. When energy efficiency of the engine increases and wasted energy decreases, fuel usage decreases at same power level. Obviously, latest high technology and selection of aircraft engine type improve the aircraft fuel efficiency. Latest trends in improving specific fuel consumption and overall efficiency show that aircraft entering aviation sector are around 80% more fuel-efficient using turboprop and high bypass ratio turbofan engines [18-22].

The overall efficiency and specific fuel consumption of aircraft engines are calculated to be using thermodynamic analysis. Unfortunately, first law analysis of thermodynamics does not show correct location information for energy destruction and losses. However, the second law must be applied to the aircraft propulsion system for determining locations and magnitudes of inefficiencies in the engine. The number of articles studied on exergy analysis and exergetic optimization of aircraft power systems has increased in the last decades [13-21, 23,24]. Especially, modeling of energetic and exergy efficiencies of turboprop engine components using neuro-evolution algorithms and hybrid genetic algorithm-artificial neural networks is studied [20, 25].

The objective of this study is to accomplish two models of artificial neural network (ANN) in predicting a business jet aircraft's exergo-emission parameters, such us exergy destruction ratio (rex,dest) and waste exergy ratio (rwex), at different flight stages and five input parameters (fuel and air mass flows, power and torque levels, gas generator shaft speed).

This study constitutes the first analysis about determining exergo-emission for aircraft power system at various flight phases. Moreover, there is no study on a detailed measure exergetic emission of turboprop power system using metaheuristics optimized machine learning modeling. Lack of this makes the paper original for aircraft energy systems and during typical flight phases.

2 System description

Turboprop engines have a great advantage for propulsive/fuel efficiency and low emissions. Turboprop engines used for regional airlines and military transport transmit the equivalent shaft power required to drive the propeller from engine turbine shaft. In this study, selected cargo aircraft (one of the most notable turboprop aircraft version is CASA/IPTN CN235) is very similar to CN235 aircraft as shown in Figure 1. Schematic of the engine and component description are shown in Figures 2 and 3, respectively.

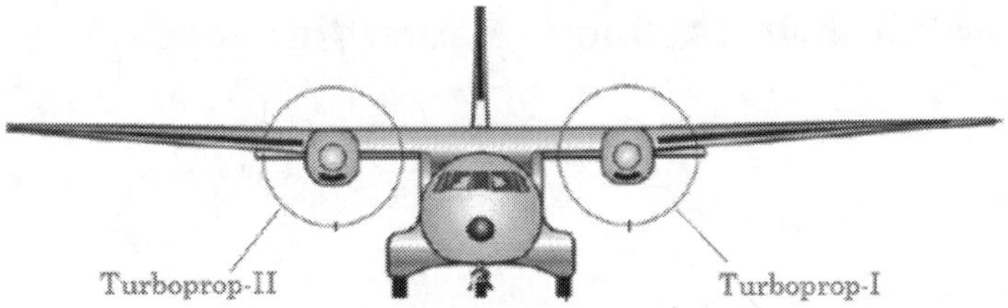

Figure 1. Turboprop engines on the cargo aircraft (modified from [20])

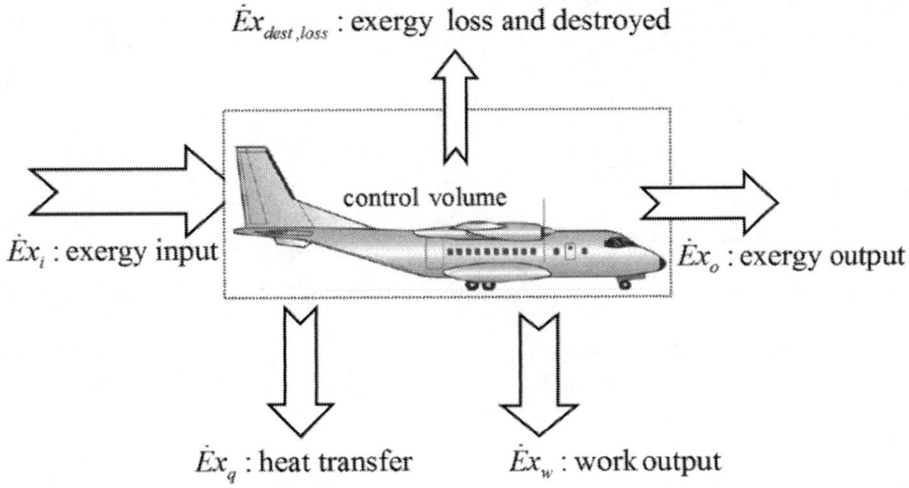

Figure 2. Exergetic control volume on the aircraft

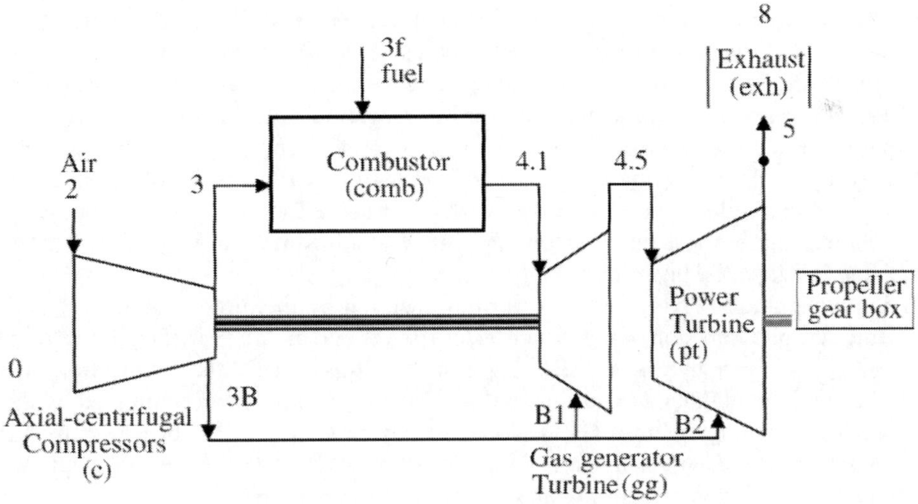

Figure 3. Main components of the turboprop power system

3 Overview of an ANN (Artificial Neural Network)

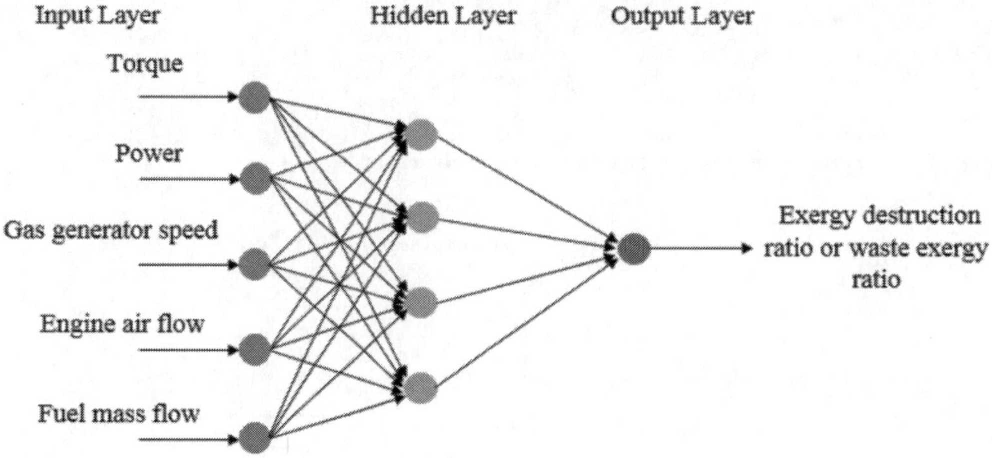

Figure 4 Architecture of the exergy destruction ratio and waste exergy ratio machine learning models

Being a type of ANN, a multilayer perceptron (MLP) (also referred to as feed-forward ANN) is composed of an input layer, an output layer and one or more hidden layers as illustrated in Figure 4.

4 GA (GENETIC ALGORITHM) Optimized BP (Back-Propagation) Feed-Forward ANNs

As proposed by a group of scientists led by Rumelhart and McCelland in 1986, the BP-ANN is a type of multilayer feed-forward network established on an error back-propagation algorithm for training in a supervised manner. There exist two stages in the learning process of a BP-ANN; namely, forward and backward propagation. Information is propagated from the input layer and then modified by the weighted connections of the hidden layer into the output layer, which enables the attainment of the input response of the ANN in the output layer. This output value given by the network is compared with the desired output. In the case of an existing error, it returns along the path and adjustment of weights and thresholds is executed in a layer by layer attitude.

Being a population driven search algorithm inspired by the biological natural selection, GA mimics the phenomenon of reproduction, crossover and mutation in the process of natural genetics. Adopting a hybrid GA-BP algorithm in this paper, the connection weights, thresholds, and the number of neurons in the hidden layer have been optimized by the GA, which provides the amalgam of GA's global search virtue and BP's local search ability in order to prevent the potential inherent problems of BP-ANNs. Moreover, GA optimization has enhanced the learning capability and accuracy of the BP-ANN.

The case study cargo aircraft turboprop engine has been tested a number of times for different power settings. The actual values of specifications related to the engine were determined from the parametric evaluation and trials and they were then used to derive the values of waste exergy ratio and actual exergy destruction ratio for this particular engine. The validation and training of the metaheuristically prepared ANN estimation models that were put forward in this study was performed with the help of actual datasets related to waste exergy ratio and exergy destruction ratio which matched to parameters of an engine still in the experimental phase.

Power (SHP), torque (N.m), speed of the gas generator (rpm), fuel mass flow (kg/s) are classified as explanatory variables, on the other hand the waste exergy ratio and exergy destruction ratio in context of 2 separate ANN models derived in this paper were classified as output variables.

In Figure 5, the lowest possible MSE's (Mean Square Error) for different rates of momentum in networks with one hidden layer can be seen whereas the same for two hidden layers is depicted in Figure 6.

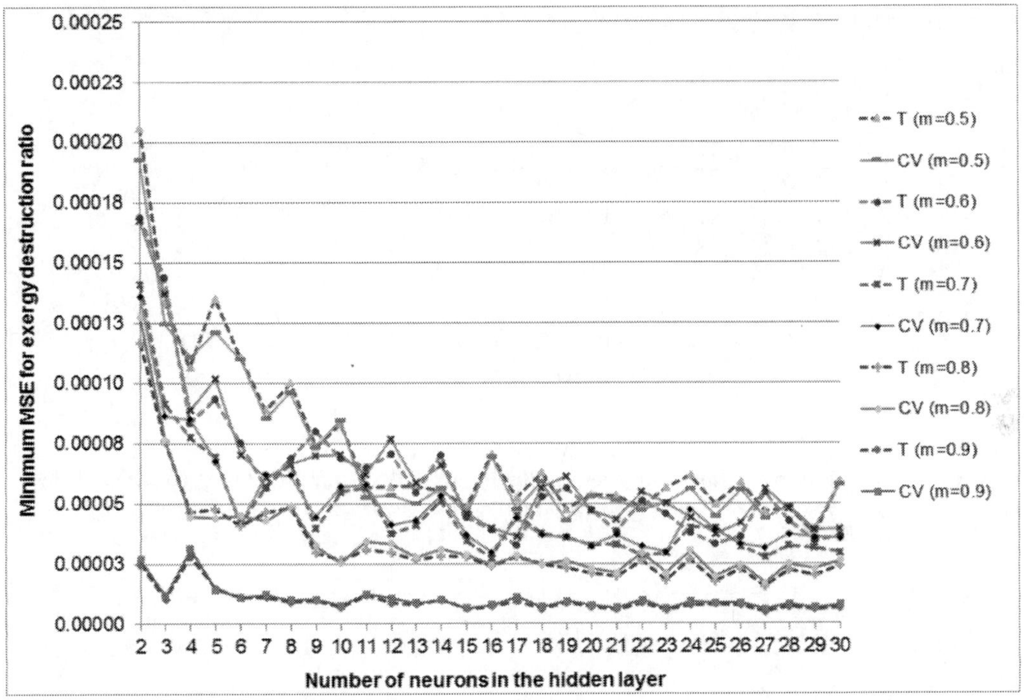

5(a)

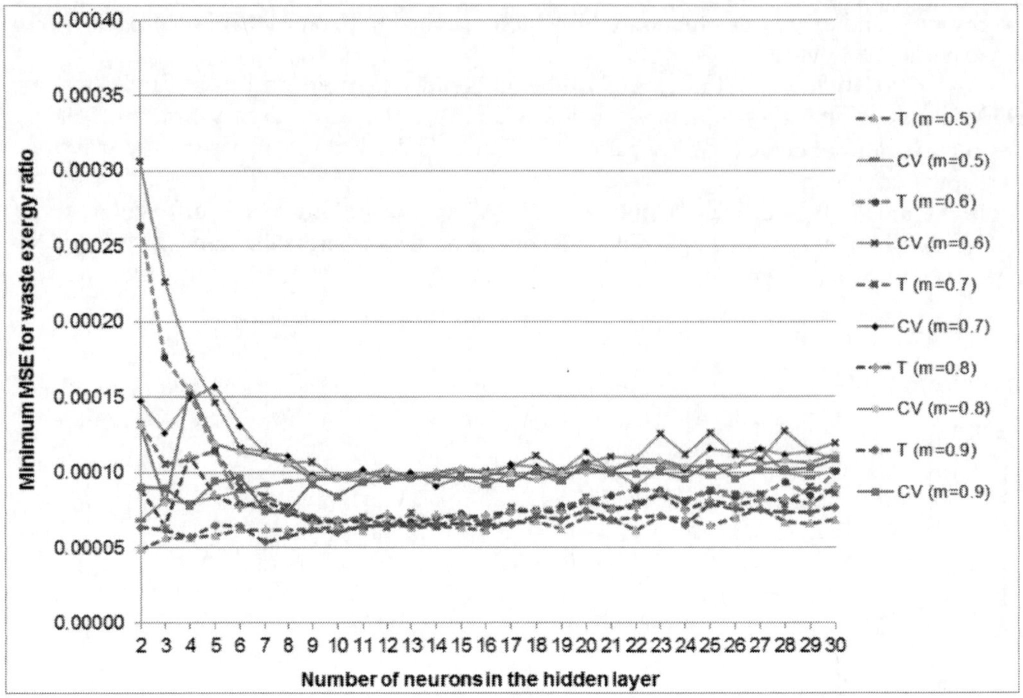

5(b)

Figure 5 Training and cross-validation results of the BP algorithm with varied momentum rates for one-hidden-layer networks: a) exergy destruction ratio, b) waste exergy ratio (T: Training, CV: Cross-validation; HL: Hidden layer)

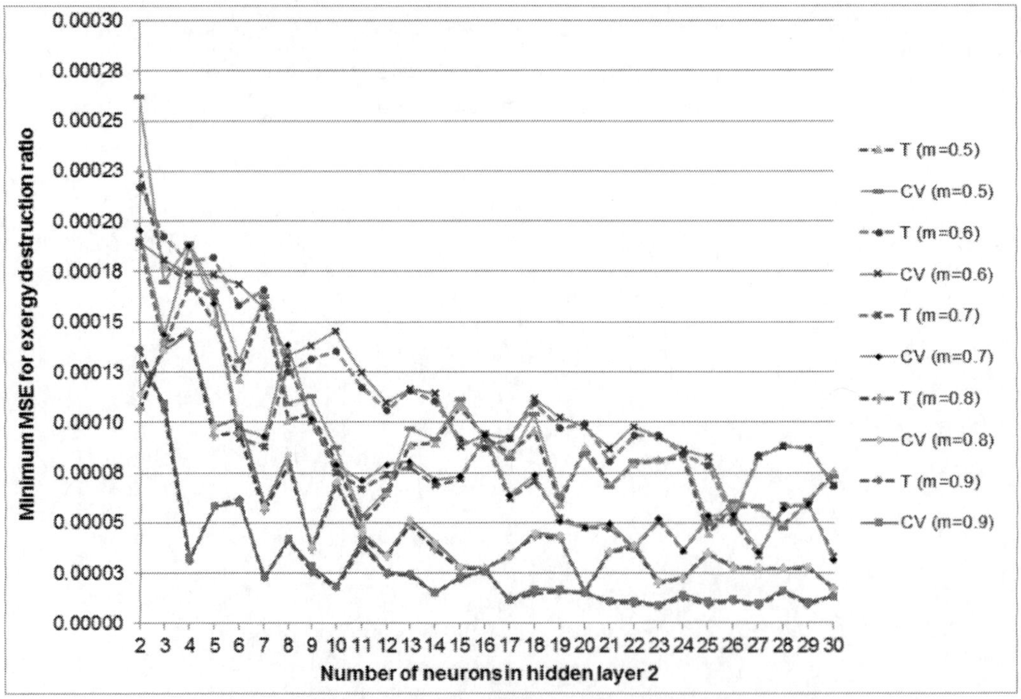

6(a)

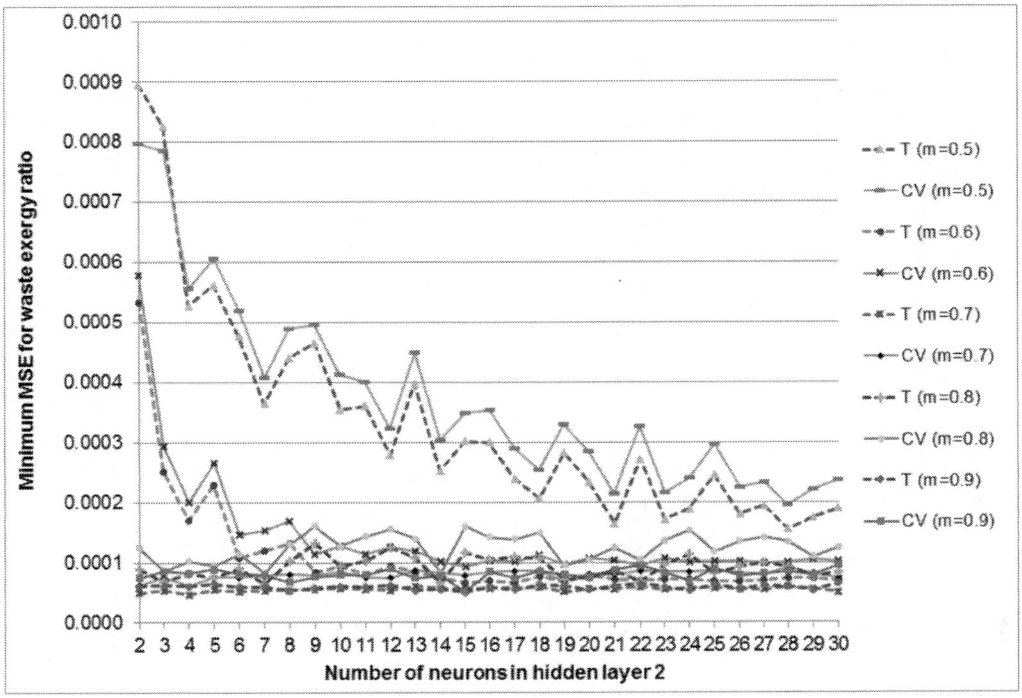

6(b)

Figure 6. Training and cross-validation results of the BP algorithm with varied momentum rates for two-hidden-layer networks: a) exergy destruction ratio, b) waste exergy ratio (T: Training, CV: Cross-validation; HL: Hidden layer)

A comparison was drawn between the values of waste exergy ratio and exergy destruction ratio which were predicted by the ideal networks and the actual values which were derived when the turboprop engine was run on the testing datasets of the two ANN models as an experiment. It can be clearly seen from Figures 7(a) and 7(b) that the actual and estimated values of the two ANN models show a close fit.

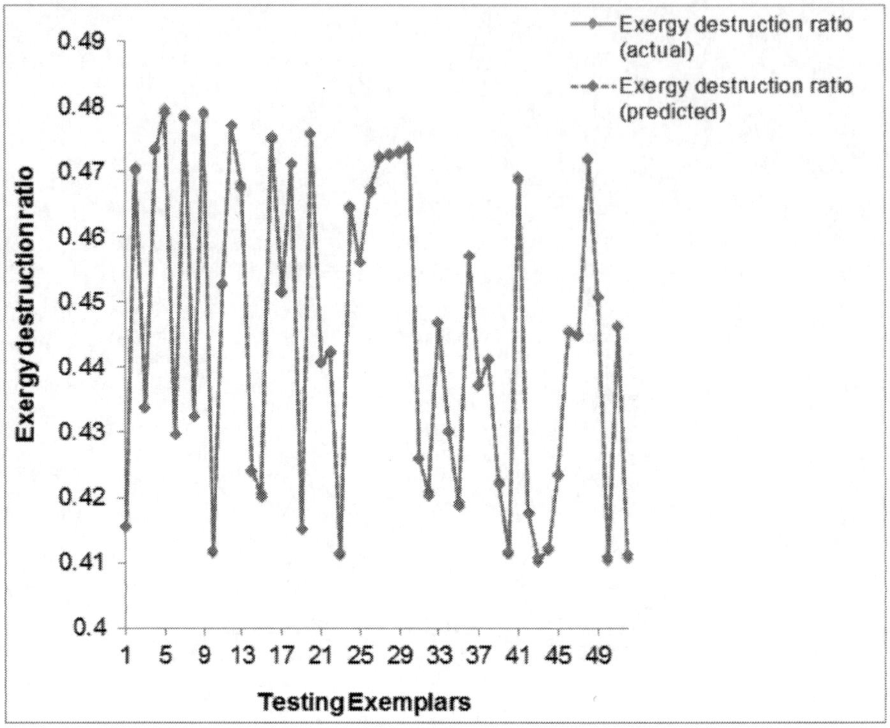

7(a)

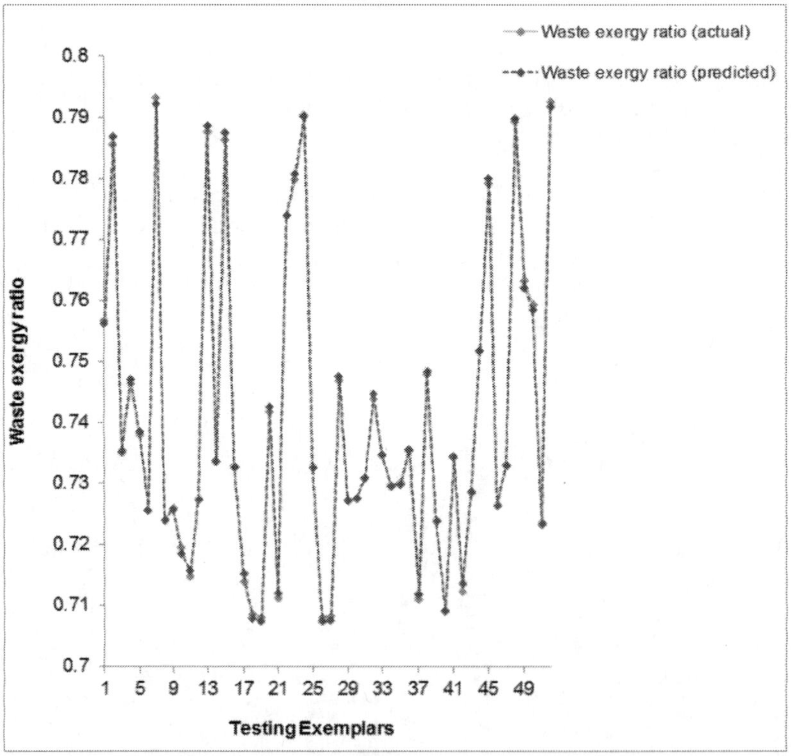

7(b)

Figure 7 Testing dataset validation by the use of optimal ANN model architectures for a) exergy destruction ratio, b) waste exergy ratio

5 Conclusions

An additional new feature of the trial design is that in the design and derivation steps actual values derived during trials and those obtained parametric evaluation of the aircraft's engine are used. Since, the effectiveness and validity of an ANN model are greatly affected by the network architecture's initial state and adapted learning algorithm hence in this never before tried method of study use of an ANN designed in a metaheuristic format further increases its precision and capability of generalization. The important results derived from modeling are described ahead:

i. For the optimum exergy destruction ratio of BP model, R value of 0.999985 is attained; whereas, in case of GA-BP model, it comes out to be 0.999987. On the other hand, the value of maximum linear correlation coefficient for waste exergy ratio came out to be 0.999851 for ANN models and 0.999869 for GA-ANN models. The fact that the R values are high and the predefined error varieties have low values point towards the fact that the accuracy of the derived ANN models is quite high in context of calculation of parameters related to exergy destruction and waste exergy.

ii. Both the efficiency and precision have showed an improvement due to application of GA-BP which is a metaheuristic design on prepared ANNs, in addition the time required for convergence was also cut short in the two models. It can be presumed that ANN is similar to a black-box like system with correctly prepared topology and initial values; this will assist in delivering the best possible output and increase the reliability of the model in context of its practical usage and precision.

iii. It was seen that the accuracy of networks with a single hidden layer was greater than that of networks with two hidden layers for exergy destruction ratio forecast model; whereas, the two-hidden-layer architectures outperformed in the case of waste exergy ratio estimation.

iv. A new benchmark of level of performance and excellence in context of high precision and results has been set up by use of metaheuristic design in estimation models that aim to determine parameters related to waste exergy and exergy destruction ratios.

v. Technological advancements in the field of aeronautics must make use of exergy concepts of artificial intelligence and metaheuristic methods so that innovations show a consistent growth and are eco-friendly as well.

Ackowledment

Authors would like to thanks Eskisehir Tehnical University in Turkey for financial and technical support. This study was supported by Eskisehir Tehnical University Scientific Research Projects Commission under the grant no: 19ADP025.

References

1. Qiang C, Ye L, Airline energy efficiency measures considering carbonabatement: A new strategic framework, Transportation Research Part D 49 246–258,(2016)
2. IATA, 2014. <http://www.iata.org/Pages/default.aspx>.
3. Lee JJ. Can we accelerate the improvement of energy efficiency in aircraft systems? Energ Convers Manage;51(1):189-96, (2010)
4. Abbas A, de Vicente J, Valero E. Aerodynamic technologies to improve aircraft performance. Aerosp Sci Technol;28(1):100-32, (2013).

5. Ponater M, Pechtl S, Sausen R, Schumann U, Huttig G. Potential of the cryoplane technology to reduce aircraft climate impact: a state-of-the-art assessment. Atmos Environ 2006;40(36):6928-44.
6. Bruno D C, Davide D F, Nicola C, Dario P, Comparative specific energy consumption between airtransport and high-speed rail transport: A practical assessment, Transportation Research Part D 52 227–243, (2017).
7. Adler, N., Pels, E., Nash, C., 2010. High-speed rail and air transport competition: game engineering as tool for cost-benefit analysis. Transp. Res. Part B 44, 812–833, (2010).
8. Hao, L., Hansen, M., Ryerson, M.S., Fueling for contingencies: the hidden cost of unpredictability in the air transportation system. Transport. Res. Part D: Transport Environ. 44, 199–210, (2016).
9. Jamin, S., Schäfer, A., Ben-Akiva, M.E., Waitz, I.A.,. Aviation emissions and abatement policies in the United States: a city-pair analysis. Transport. Res. Part D: Transport Environ. 9 (4).295-317, (2004).
10. Yang, H., Zhang, A.,. Effects of high-speed rail and air transport competition on prices, profits and welfare. Transp. Res. Part B 46 (10), 1322–1333, (2012).
11. Sahin, O., Turan, O.,. Evaluation of aircraft descent profile. Energy Procedia 95, 308–313, (2016).
12. Meric, O.S., Optimum arrival routes for flight efficiency. J. Power Energy Eng. 3, 449–452, (2015).
13. Aydin, H., Turan, O., Karakoc, T.H., Midilli, A., Exergo-sustainability indicators of a turboprop aircraft for the phases of a flight. Energy 58, 550–560, (2013).
14. Atilgan, R., Turan, O., Altuntas, O., Aydin, H., Synylo, K., Environmental impact assessment of a turboprop engine with the aid of exergy. Energy 58, 664–671, (2013).
15. Turan, O., Aydin, H.,. Exergetic and exergo-economic analyses of an aero-derivative gas turbine engine. Energy 74, 638–650, (2014).
16. Turan, O.,. Effect of reference altitudes for a turbofan engine with the aid of specific-exergy based method. Int. J. Exergy 11, 252–270, (2012).
17. Turan, O., An exergy way to quantify sustainability metrics for a high bypass turbofan engine. Energy 86, 722–736, (2015).
18. Aydin H, Turan O, Karakoc TH, Midilli A. Sustainability assessment of PW6000 turbofan engine:

an exergetic approach. International Journal of Exergy 14(3): 388-412, (2014).

19. Aydin H, Turan O, Karakoc TH, Midilli A. Exergetic sustainability indicators as a tool in commercial aircraft: a case study for a turbofan engine. Int J Green Energy, 12: 28-40, (2015)
20. Baklacioglu T, Turan O, Aydin H. Metaheuristic approach for an artificial neural network: Exergetic
sustainability and environmental effect of a business aircraft, Transport. Res. Part D: Transport Environ, 63 445–465, (2018).
21. Aydin H, Turan O. Numerical calculation of energy and exergy flows of a turboshaft engine for power generation and helicopter applications. Energy,; 115 (1): 914-23, (2016).
22. Cilgin, ME, Turan, O. Entropy generation calculation of a turbofan engine: a case of CFM56-7B, Int J Turbo Jet Eng; 58: 664-71, (2017).
23. Turan, O., Aydin, H.,. Exergetic and exergo-economic analyses of an aero-derivative gas turbine engine. Energy 74, 638–650, (2014).
24. Turan, O., Exergetic effects of some design parameters on the small turbojet engine for unmanned air vehicle applications. Energy 46, 51–61, (2012).
25. Baklacioglu T, Turan O, Aydin H. Metaheuristics Optimized Machine Learning Modelling of Environmental Exergo-Emissions for an Aero-Engine, Int J Turbo Jet Eng; https://doi.org/10.1515/tjj-2019-0037, (2019).

Horizontal flight trajectory optimization considering RTA constraints

Alejandro Murrieta-Mendoza[1,*], *Hugo* Ruiz[2], and *Ruxandra* Mihaela Botez[3]
[1]LARCASE, École de Technologie Supérieure, Université du Québec, Canada
[2]LARCASE, École de Technologie Supérieure, Université du Québec, Canada
[3]LARCASE, École de Technologie Supérieure, Université du Québec, Canada

Abstract. The increasing of flights around the world has led to various problems for the aeronautical industry such as saturated air space and higher levels of fossil fuel consumption. The way in which en-route flights are handled should be improved in order to increase airways' capacity. A solution is to make aircraft to arrive at specific waypoints at a time constraint called Required Time of Arrival (RTA). Fossil fuel brings as a consequence the release of polluting particles to the atmosphere such as carbon dioxide and nitrogen oxides. It is thus desirable to compute the most economical trajectory in terms of fuel burn while fulfilling the RTA constraint. This article proposes a horizontal reference trajectory optimization algorithm based on the Particle Swarm Optimization technique in order to reduce fuel burn while fulfilling the RTA constraint. Results showed that for a flight without RTA constraint, up to 4% of fuel can be saved comparing against the trajectory of reference. The algorithm was normally able to meet the RTA constrain. However, aggressive RTA constraints might reduce the optimization levels of fuel compared with flights without RTA constraint.

1 Introduction

Air traffic is expected to grow in the forthcoming years according with market outlooks by Airbus1 and Boeing2. The geographical places where traffic is expected to have an important growth are those in development: Asia, and Latin America.

The traffic increase has a direct impact on the fossil fuel required to power flights, and thus in the number of polluting particles released to the atmosphere. Among these particles, Carbon dioxide (CO2), Nitrogen Oxides (NOx), and hydrocarbons (HC) are of interest: CO2 has been of special interest due to its effect on the global warming effect, NOx has been attributed to deplete the ozone layer [1], and HC cause health problems and reduces the air quality [2].

From an economical perspective, it is expected that government will impose economic measures for pollution avoidance, especially for CO2. Fossil fuel also represents an important expense for airlines as any reduction on fuel burn could positively impact profit.

[*] Corresponding author: aj70290@ens.etsmtl.ca

Because of these two reasons (pollution and economic consequences), the importance of reducing fuel consumption becomes evident.

By recognizing the impact of fuel to profits, airlines have implemented different measurements to reduce fuel consumption. As discussed in [3], these measurements consists, among others, in improving efficiency by washing dirt off, reduce the use of the auxiliary power, and the aircraft weight reduction by using lighter equipment.

Flight operations might also reduce fuel consumption. According to different studies, in the continental United States airspace [4, 5] and in the Turkish airspace [6] aircraft do not fly at their optimal speeds and altitudes. It is thus of interest to provide the aircraft with a more economical reference trajectory.

One of the first improvements in trajectory optimization was in the descent phase. A descent procedure called a Continuous Descent Approach (CDA) was developed. Nevertheless, care must be given to correctly compute the beginning of descent location point to avoid missing the descent leading to a missed approach, which according to [7] it is expensive in terms of fuel consumption. The CDA technique has been implemented at different airports and gave good results [8-11].

Reference trajectory and airspace optimization are of interest to reduce fuel consumption, and to increment the airspace capacity while guaranteeing aircraft separation and safety. Different countries have been working in various programs to improve and optimize the airspace. In this context, the United States are developing the Next Generation of ATM System (NextGen), and Europe has been working with the Single European Sky ATM Research (SESAR).

These programs proposed the Trajectory Based Operations (TBO) or the Intended Based Operations (IBO). By use of these operations, the aircraft would be able to negotiate its optimal reference trajectory with traffic control instead of having an imposed reference trajectory. Future systems intend to use intended 4D trajectories, which means that aircraft position (lat/lon), speed, altitude and Required Time of Arrival (RTA) to a given (or to a set) waypoint will be required.

Reference trajectory optimization could be classified in two main different types: the set of coordinates (waypoints) an aircraft has to follow, which is known as Lateral NAVigation reference trajectory (LNAV) and the set of speeds and altitudes that the aircraft has to follow, which is known as Vertical NAVigation reference trajectory (VNAV). In VNAV, the aircraft can perform changes in altitude (step climbs) in order to find favourable weather or less air resistance (drag force) into the flight in order to reduce fuel consumption [12].

Computing the most efficient trajectory in terms of fuel burn is difficult since the aircraft fuel burn is affected by different parameters such as weather, the change of mass due to fuel consumption, the diversity of combinations available to perform the flight, and the infinite positions where to execute a step-climb.

To compute the most economical trajectory, different algorithms using optimal control have been used as in [13-18] have been implemented. The VNAV reference trajectory for all flight phases has been optimized by using the golden search section and step climbs [19], a search space reduction algorithms were developed in [20, 21], and the branch and bound algorithm was implemented in [22], both algorithms were later coupled to reduce the computation time [23]. Genetic Algorithms were also used to find the waypoints where to maximize fuel savings [24].

For the LNAV reference trajectory Classical optimization techniques such as finding the shortest path technique have been used to find the most efficient LNAV trajectory [25, 26] a as well as avoiding obstacles in LNAV for general purposes aircraft. [27]. Dijkstra's Metaheuristic algorithms have been successfully used to compute the optimal LNAV trajectory at fixed altitudes [28-30].

The objective of this paper is to validate the potential of the Particle Swarm Optimization (PSO) Algorithm in order to find the most economical reference trajectory in LNAV between two points for intercontinental flights. Through this paper, the word "optimal" refers to the trajectory that provides the most economical fuel consumption. This paper is arranged as follows. Firstly, the fuel burn model is described along with the way in which the flight cost is computed. Next, the weather model is explained. Then, the search space and the objective function are presented. Then, the PSO algorithm implementation is detailed, followed by results and conclusions.

2 Methodology

2.1 Fuel Burn Model and Trajectory Cost Computation

Captions The fuel burn is computed using a Performance DataBase (PDB) which was developed using flight test data. The PDB is a set of tables that provide for given inputs a certain output. The inputs for the cruise phase table are the Aircraft weight, Speed (Mach or IAS), altitude, and the International Standard Atmosphere Temperature Deviation (ISA_DEV). The ISA_DEV gives the difference between the real temperature and the temperature provided by the standard atmosphere model The PDB gives as result the fuel flow for those flight conditions.

Use of a PDB requires a procedure to compute the flight fuel consumption. A complete methodology to compute the flight cost from a database by considering all flight phases such as climb, acceleration, cruise, descent, and deceleration was discussed in [29, 30]. In this paper, only the procedure to compute the fuel consumption in the cruise phase is briefly explained.

Because of the fact that not each required input value is available to provide the required output (fuel flow for the cruise phase), interpolations are required. For these interpolations to be performed, first the lower and the upper bounds surrounding the required values are identified, which is done for each input. A set of linear interpolations are executed to compute the required output from the identified inputs. Figure 1 provides a scheme of the interpolations required to obtain a given fuel flow for a flight condition.

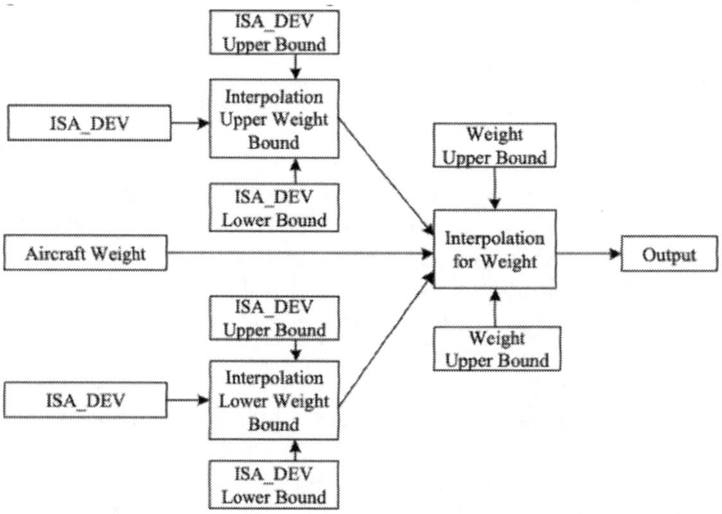

Figure. 1. Interpolation Schema

In order to obtain the fuel burn for a given cruise distance, the trajectory is divided in different segments. The obtained fuel flow is multiplied by the flight time required to travel a given segment as shown in Eq. (1).

$$FB_{seg} = FF_{seg} * Flight\ Time \quad (1)$$

As fuel is burned during flight, the aircraft mass diminishes over time. This fuel burned should be reduced from the total aircraft weight every time the aircraft flies a pre-defined distance. This distance is important since the accuracy of results depends on how often the aircraft mass is updated, the more often the mass is updated, the more accurate the fuel consumption computation is. However, updating the aircraft mass too often will have as a consequence the execution of a high number of interpolations, that might increase the computation time. These computations are performed until the last segment is flown.

The flight time to travel a given distance must be computed using the Ground Speed (GS), which takes into account wind effects as shown in Eq (2), not the True Air Speed (TAS) or the Indicated Mach number.

$$GS = TAS \pm Wind\ Speed \quad (2)$$

2.2 Weather Data

Weather data was obtained from the open weather data forecast generated daily by Weather Canada. These data are provided under the form of a grid that contains the weather information for all over the world for different altitudes at different hours. This grid is provided with a resolution of 0.6 km x 0.6 km.

A detailed method that process the weather information and computes the ground speed can be found in [31]. A brief description on the way to obtain the weather information is described. Firstly, the 4 points of the grid surrounding the aircraft are identified. Then, the four points are interpolated for the required altitude and time. Finally, a bilinear interpolation can be executed to obtain the required weather parameter (e.g. Wind, Temperature) at the aircraft position, and time. Figure 2 schematically shows the steps, and the interpolations followed to compute the weather information before executing bi-linear interpolations.

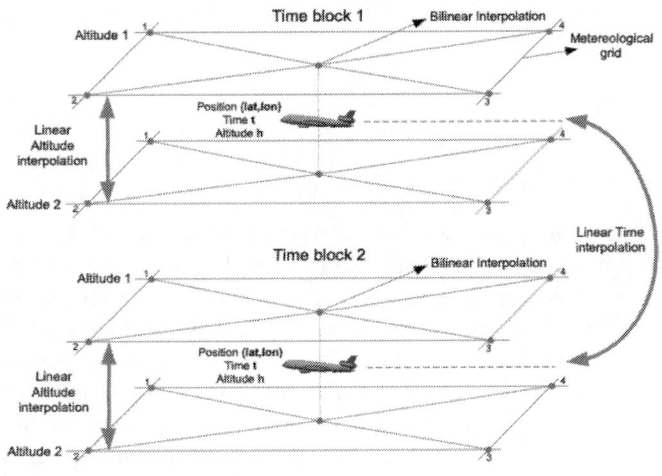

Figure. 2. Weather Interpolations

2.3 Search Space

The search space can be modelled as a weighted graph where edges (E) connect nodes (V). As the study of interest concerns the cruise phase, the first waypoint is the Top of Climb (TOC) and the last waypoint is the Top of Descent (TOD).

In Figure 3, every "node" is considered as a possible waypoint where the aircraft can fly to. The weight expressed in every vertex corresponds to the cost required to reach the connected node.

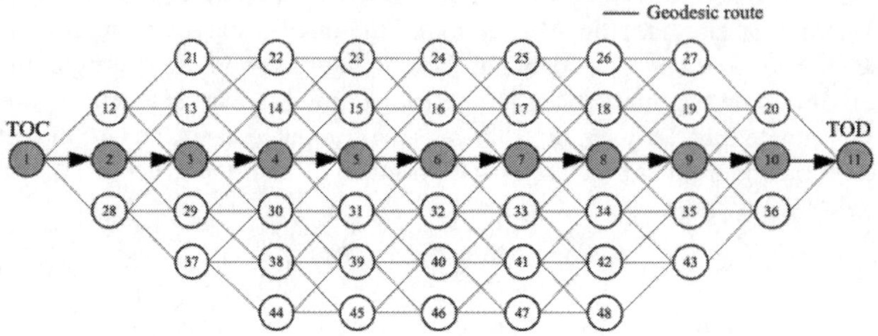

Figure. 3. Lateral Reference Trajectory Asymmetric Graph

2.4 The Optimal Solution and the Objective Function

The algorithm aims to minimize Eq (3).

$$Flight\ Cost = Fuel\ Burn + abs(RTA\_error) * RTA\_cost \quad (3)$$

where RTA_error is the difference between the arrival time and the RTA constraint. RTA_cost is the unit cost of missing the RTA. A flight that misses the RTA constraint becomes more expensive, as it is penalized with an extra cost (RTA_cost), thus it is most likely to be rejected by the algorithm. This cost increase is given by the time difference between the arrival time, and the RTA constraint multiplied by a user pre-defined RTA_cost. The larger the difference between the arrival time and the required RTA is, the more expensive the flight cost becomes. It should be noted that this artificial flight cost is an artificial cost used only to reject trajectories that do not fulfil the RTA constraint.".

3 The Algorithm

3.1 The Particle Optimization Algorithm

The objective is to find the combinations of waypoints from the ToC to the ToD that would provide the most economical way to connect them. For this is a combinatory problem resolution, it has been proved that the metaheuristic PSO algorithm performs well in finding very good solutions. The PSO, first described in [32], mimics the behaviour of social animals which follow a leader during their activities such as flocks of birds or fish schools.

The PSO algorithm is iterative. At every iteration, particles are modified in order to find new solutions within the search space. These particles are influenced by their own position, a global leader's positions and another particle's positions. Each particle in the PSO algorithm can be define as a set of waypoints from ToC to ToD, or in other words, as a complete trajectory. The equations that model these influences are shown in Eq (4) and Eq (5).

$$D_i(k+1) = \omega R_\omega D_i(k) + c_1 R_1 (B_l - X_i(k)) + c_2 R_2 \left(B_g - X_i(k)\right) \tag{4}$$

$$X_i(k+1) = X_i(k) + D_i(k+1) \tag{5}$$

where i is a trajectory, and k is the current iteration. D is the particle's displacement associated with the parameter i, which indicates if the displacement is in the lateral or vertical reference trajectory. X is the next particle's location. B_g is the best global trajectory position, and B_l is the best local trajectory position. In order to have an influence in the solution convergence, there are 3 random parameters R_ω, R_1 and R_2 with values between 0 and 1. The role of these parameters R is to determine the influence on the particle's movement. The parameters c_1 and c_2 are positive constants provided by the user, which influence the particle movement in reference to the B_g, B_l. The parameter ω refers to the aircraft's inertia and it is set by user. The inertia represents the particles tendency to keep the same direction on the search space.

3.2 The LNAV Optimization Algorithm

The steps required to compute the optimal LNAV reference trajectory is as follows

Step 1: Random Trajectory Generation. Different clusters with a pre-defined number of trajectories are randomly generated.

Step 2: Flight Cost Evaluation and Selection. All trajectories are evaluated. The most economical is set as the current global leader. The most economical trajectory of each cluster is defined as a local leader.

Step 3: Random Vectors Generation. R vectors are randomly generated.

Step 4: Displacement Computation. The displacement for every single trajectory is computed with Eq (4) and the vectors generated in Step 3. Displacement for this work consists on the set of waypoints to follow and the Mach Speed.

Step 5: New position Computation. The new particle's position (reference trajectory) is updated with the computed displacement in Step 4 using Eq (5).

Step 6: New trajectories evaluation and leaders updated. The flight cost of every trajectory is evaluated. The most economical trajectory is defined as the global leader particle and the most economical trajectory of every cluster is defined as a local leader. Step 7: Stopping criteria. The algorithm stops if a given number of iterations is reached. Otherwise, it returns to step 3.

4 Results

The fuel saving potential of the trajectories provide by the algorithm was evaluated for a two engine, wide body aircraft with a mass of 125000 kg at the ToC and a fixed cruise altitude of 34,000 ft. It was considered that the aircraft is flying under the TBO concept, free to fly anywhere in the search space. All flights were intercontinental flights. Optimized flights are compared with their geodesic trajectories (shortest paths in a sphere).

Two sets of LNAV result are presented, one considering only the fuel optimization potential and the other one the optimization with a RTA constraint.

4.1 LNAV Optimization

The first set of results aim to validate the potential of the PSO algorithm to reduce fuel burn. Six different flights: Montreal to Innsbruck (YUL – INS), Montreal to Grand Canary Island (YUL – LEP), Montreal to London (YUL – LON), Vancouver to Grand Canary Island (YVR – LPS), and Los Angeles to Innsbruck (LAX – INS). Fuel saving results are shown in Figure 4.

It can be seen from Figure 4 that all flights gave better results than the geodesic trajectory of reference. It can also be seen that the longest trajectories (YVR – LPS and LAX – INS) gave more fuel savings than the other routes. This is because the longer the trajectories the algorithm was, depending the case, able to benefit from tail winds or avoid head winds for a longer period.

4.2 LNAV Optimization with RTA constraint

The second set of results aimed to observe the optimization savings when a RTA constraint was set. In this case, the RTA allowed error was of +/- 30 seconds, and the Mach numbers that the algorithm could select were between 0.62 and 0.8. The same flight cases were evaluated as the ones in Section 4.1. The RTA constrain was selected quit aggressive by setting it to be the same value as the reference trajectory. Fuel saving results are shown in Figure 5.

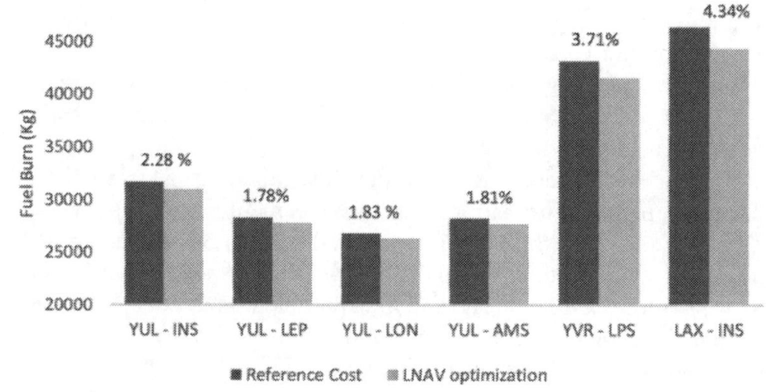

Figure. 4. Flight Savings without a RTA Constraint.

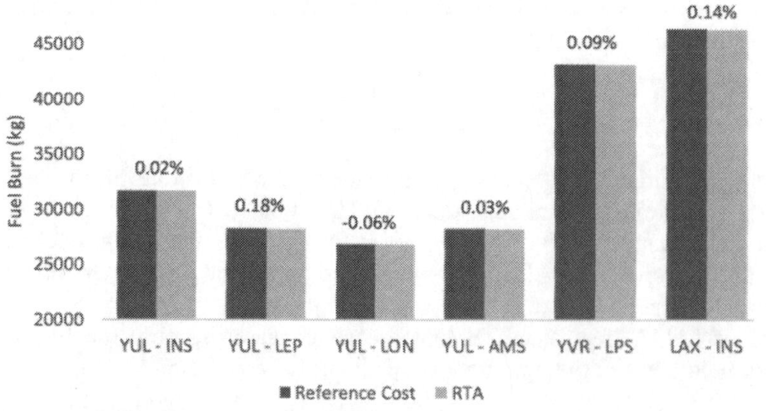

Figure. 5. Flight Savings with a RTA Constraint.

It can be seen from Figure 5 that by imposing an aggressive RTA constraint, the optimization is reduced to practically zero. Even the flight YUL – LON recorded a more expensive flight than the reference flight.

The RTA constraint information is shown in Table 1 where in almost all cases, the RTA constraint was met. The only exception were the flights YVR – LPR and LAX – INS which are the longest ones. It can be seen that the first 4 flights (with similar distances) were all close to the RTA limit. The highest error for the RTA was given by the LAX – INS flight which was also the longest flight. This result was due to the fact that the algorithm only generates one Mach number during flight. To reduce this error, the algorithm should be able to change aircraft speed during flight.

Table 1. RTA performance.

Flight	TOC Time (h)	TOD RTA (h)	Optimal Arrival (h)	Error (s)	Flight Time (h)
YUL - INS	4	11.742	11.741	28	7.7
YUL - LEP	4	10.812	10.813	25	6.8
YUL - LON	4	10.492	10.495	23	6.5
YUL - AMS	4	10.840	10.838	25	6.8
YVR - LPS	4	14.745	14.736	39	10.7
LAX - INS	4	15.707	15.695	42	11.7

5 Conclusion

The LNAV optimization algorithm based on PSO was able to find combinations of waypoints and a Mach number that would reduce the flight consumption up to 4% with respect to the reference trajectory. This fact could potentially lead to a reduction of polluting emissions and the flight cost.

The algorithm should be able to generate more than one Mach number per flight in order to meet the RTA constraint and to save fuel for long-haul flights. It was observed that by imposing aggressive RTAs, the optimization opportunities might be reduced.

Acknowledgements

This research was conducted at the Laboratory of Applied Research in Active Controls, Avionics and Aeroservoelasticity (LARCASE) within the frame of the global project "Optimized Descent and Cruise" with funds from the Business-led Network of Centers of Excellence Green Aviation Research & Development Network (GARDN). For more information please visit http://larcase.etsmtl.ca. The authors would like to thank Rex Haygate, Dominique Labour and Yvan Blondeau from CMC- Electronics – Esterline, and Oscar Carranza from LARCASE. The authors would also like to thank CONACYT in Mexico and the FQRNT in Quebec, Canada.

References

1. P. J. Crutzen. "The influence of nitrogen oxides on the atmospheric ozone content," *Q.J.R. Meteorol. Soc* **Vol**. 96, No. 408, (1970), pp. 320-325. doi: 10.1002/qj.49709640815
2. E. Robinson. "Hydrocarbons in the atmosphere," *Pure and applied geophysics* **Vol**. 116, No. 2-3, (1978), pp. 372-384. doi: 10.1007/bf01636892
3. D. McConnachie, C. Wollersheim, and R. J. Hansman. "The Impact of Fuel Price on Airline Fuel Efficiency and Operations," *2013 Aviation Technology, Integration, and Operations Conference*. American Institute of Aeronautics and Astronautics, (2013).
4. L. Jensen, J. R. Hansman, J. Venuti, and T. Reynolds. "Commercial Airline Altitude Optimization Strategies for Reduced Cruise Fuel Consumption," *14th AIAA Aviation Technology, Integration, and Operations Conference*. American Institute of Aeronautics and Astronautics, (2014).
5. L. Jensen, J. R. Hansman, J. C. Venuti, and T. Reynolds. "Commercial Airline Speed Optimization Strategies for Reduced Cruise Fuel Consumption," *2013 Aviation Technology, Integration, and Operations Conference*. American Institute of Aeronautics and Astronautics, Los Angeles, USA, (2013).
6. E. T. Turgut, M. Cavcar, O. Usanmaz, A. O. Canarslanlar, T. Dogeroglu, K. Armutlu, and O. D. Yay. "Fuel flow analysis for the cruise phase of commercial aircraft on domestic routes," *Aerospace Science and Technology* **Vol**. 37, No. 0, (2014), pp. 1-9. doi: http://dx.doi.org/10.1016/j.ast.2014.04.012
7. A. Murrieta-Mendoza, R. M. Botez, and S. Ford. "New method to compute the missed approach fuel consumption and its emissions," *The Aeronautical Journal* **Vol**. 120, No. 1228, (2016), p. 18. doi: http://dx.doi.org/10.1017/aer.2016.37
8. T. Kwok-On, B. Daniel, and W. Anthony. "Development of Continuous Descent Arrival (CDA) Procedures for Dual-Runway Operations at Houston Intercontinental," *6th AIAA Aviation Technology, Integration and Operations Conference (ATIO)*. American Institute of Aeronautics and Astronautics, (2006).
9. I. De Lépinay, D. Dimitriu, and A. Melrose. "Environmental impact assessment of Continuous Descent Approaches at Manchester and Bucharest airports," *35th International Congress and Exposition on Noise Control Engineering, INTER- NOISE 2006*. Vol. 6, Institute of Noise Control Engineering of the USA -, Honolulu, HI, United states, (2006), pp. 3856-3863.
10. T. Kwok-On, W. Anthony, and B. John. "Continuous Descent Approach Procedure Development for Noise Abatement Tests at Louisville International Airport, KY," *AIAA's 3rd Annual Aviation Technology, Integration, and Operations (ATIO) Forum*. American Institute of Aeronautics and Astronautics, (2003).
11. D. Novak, T. Buckai, and T. Dadisic. "Development, Design and Flight Test Evaluation of Continuous Descent Approach Procedure in FIR Zagreb," *Scientific Journal on Traffic and Transportation Research* **Vol**. 21, No. 5, (2009), pp. 319- 329. doi: http://dx.doi.org/10.7307/ptt.v21i5.247
12. J. A. Lovegren. "Estimation of potential aircraft fuel burn reduction in cruise via speed and altitude optimization strategies." Massachusetts Institute of Technology (2011), p. 97.

13. P. Bonami, A. Olivares, M. Soler, and E. Staffetti. "Multiphase Mixed-Integer Optimal Control Approach to Aircraft Trajectory Optimization," *Journal of Guidance, Control, and Dynamics* **Vol**. 36, No. 5, (2014), pp. 1267-1277. doi: http://dx.doi.org/10.2514/1.60492

14. J. A. Cobano, D. Alejo, G. Heredia, and A. Ollero. "4D Trajectory Planning in ATM With an Anytime Stochastic Approach," *Proceedings of the 3rd International Conference on Application and Theory of Automation in Command and Control Systems*. ACM, Naples, Italy, (2013).

15. B. Korn, H. Helmke, and A. Kuenz. "4D Trajectory Managemetn In The Extended TMA: Coupling AMAN And 4D FMS For Optimized Approach Trajectoreis," *25th Congress of International Council of the Aeronautical Sciences*. Vol. Paper ICAS 2006-8.8.2 Optimage Ltd, Hamburg, Germany, (2006).

16. M. Soler-Arnedo, M. Hansen, and B. Zou. "Contrail Sensitive 4D Trajectory Planning with Flight Level Allocation Using Multiphase Mixed-Integer Optimal Control," *AIAA Guidance, Navigation, and Control (GNC) Conference*. American Institute of Aeronautics and Astronautics, (2013).

17. C. Tsotskas, T. Kipouros, and M. Savill. "Biobjective Optimisation of Preliminary Aircraft Trajectories," *Evolutionary Multi-Criterion Optimization*. Vol. 7811, Springer Berlin Heidelberg, (2013), pp. 741-755.

18. A. Valenzuela, and D. Rivas. "Optimization of Aircraft Cruise Procedures Using Discrete Trajectory Patterns," *Journal of Aircraft* **Vol**. 51, No. 5, (2014), pp. 1632-1640. doi: http://dx.doi.org/10.2514/1.C032041

19. R. S. Félix-Patrón, R. M. Botez, and D. Labour. "New Altitude Optimisation Algorithm for the Flight Management System CMA-9000 Improvement on the A310 and L-1011 Aircraft," *The Aeronautical Journal* **Vol**. 117, No. 1194, (2013), pp. 787-805.

20. J. Gagné, A. Murrieta-Mendoza, R. Botez, and D. Labour. "New Method for Aircraft Fuel Saving Using Flight Management System and Its Validation on the L- 1011 aircraft," *2013 Aviation Technology, Integration, and Operations Conference*, (2013). doi: http://dx.doi.org/10.2514/6.2013-4290

21. A. Murrieta-Mendoza, and R. M. Botez. "Vertical Navigation Trajectory Optimization Algorithm For A Commercial Aircraft," *AIAA/3AF Aircraft Noise and Emissions Reduction Symposium*, (2014). doi: http://dx.doi.org/10.2514/6.2014- 3019

22. A. Murrieta-Mendoza, B. Beuze, L. Ternisien, and R. M. Botez. "New reference trajectory optimization algorithm for a flight management system inspired in beam search," *Chinese Journal of Aeronautics*, (2017). doi: http://dx.doi.org/10.1016/j.cja.2017.06.006

23. A. Murrieta-Mendoza, L. Ternisien, B. Beuze, and R. M. Botez. "Aircraft Vertical Route Optimization by Beam Search and Initial Search Space Reduction," *Journal of Aerospace Information Systems* **Vol**. 15, No. 3, (2018), pp. 157-171.doi: http://dx.doi.org/10.2514/1.i010561

24. A. Murrieta-Mendoza, R. S. Félix-Patrón, and R. M. Botez. "Flight Altitude Optimization Using Genetic Algorithms Considering Climb and Descent Costs in Cruise with Flight Plan Information," *SAE 2015 AeroTech Congress & Exhibition*. SAE International, Seattle, USA, (2015), p. 9.

25. A. Murrieta-Mendoza, and R. M. Botez. "Lateral Navigation Optimization Considering Winds And Temperatures For Fixed Altitude Cruise Using The Dijkstra's Algorithm," *International Mechanical Engineering Congress & Exposition*. Montreal, Canada, (2014).

26. A. Murrieta-Mendoza, C. Romain, and R. M. Botez. "Commercial Aircraft Lateral Flight Reference Trajectory Optimization," *20th IFAC Symposium on Automatic Control in Aerospace ACA 2016* **Vol.** 49, No. 17, (2016), pp. 1-6. doi: http://dx.doi.org/10.1016/j.ifacol.2016.09.001
27. E. Rippel, A. Bar-Gill, and N. Shimkin. "Fast Graph-Search Algorithms for General-Aviation Flight Trajectory Generation," *Journal of Guidance, Control, and Dynamics* **Vol.** 28, No. 4, (2005), pp. 801-811.doi: http://dx.doi.org/10.2514/1.7370
28. R. S. Félix-Patrón, A. Kessaci, and R. Botez. "Horizontal Flight Trajectories Optimisation for Commercial Aircraft Through a Flight Management System " *The Aeronautical Journal* **Vol.** 118, No. 1210, (2014), p. 20.
29. A. Murrieta Mendoza, A. Bunel, and R. Botez. "Aircraft Lateral Flight Optimization Using Artificial Bees Colony," *International Conference on Air Transport INAIR 2015*. Amsterdam, the Netherlands, (2015).
30. A. Murrieta-Mendoza, A. Hamy, and R. M. Botez. "Lateral Reference Trajectory Algorithm Using Ant Colony Optimization," *16th AIAA Aviation Technology, Integration, and Operations Conference*. American Institute of Aeronautics and Astronautics, Washington, D.C., (2016).
31. A. Murrieta-Mendoza. "Vertical and Lateral Flight Optimization Algorithm and Missed Approach Cost Calculation.." Vol. Master, École de Technologie Supérieure, Montreal, (2013), p. 114.
32. J. Kennedy and R. Eberhart, "Particle swarm optimization," Proceedings of ICNN'95 - International Conference on Neural Networks, Perth, WA, Australia, 1995, pp. 1942-1948 **Vol.**4. doi: http://10.1109/ICNN.1995.488968

Exergo-economic cost accounting for PW4000 turbofan engine and its components

Hakan Aygun[1] , *Mehmet E.* Cilgin[1], and *Onder* Turan[1,*]
[1]Eskisehir Technical University, Faculty of Aeronautics and Astronautics, Eskisehir, Turkey

Abstract. You The several series of PW4000 high bypass turbofan engine have used so far in many aircrafts. These commercial engines have played a crucial role on passenger and freight transportations. Namely, these engines are closely related to the environment impacts and security of energy supply. In this article, exergoeconomic analysis which is useful tool to investigate existing potential for improvement of the a system efficiency were carried out. The assesment, design and optimization of energy consuming systems are performed by means of these analyses. Therefore, thermo-economic costs were assigned to existing exergetic values of PW400 engine. Also exergo-economic performance parameters were evaluated. Finally, exergo-economic deputy parameters were examined to understand relations with exergo-economic parameters. Based on the results of exergo-economics analysis, for Fan and exhaust, specific thrust costs are estimated 5.7051 $/hkN and 68.45$/hkN respectively. Also exergo-economics factor of PW4000 is found 7.958 % , while relative cost difference is determined at highest rate with 24.458 % for combustion chamber . With examination relations between economic variables and exergo-economic performance parameters, the change between 0.6 and 1.2 $/kg in the fuel price leads to increase the exhaust and fan specific thrust costs with 82.4701 $/hkN and 5.4332 $/hkN respectively. It is expected that conclusions of this study are helpful to notify exergo-economic impact of PW4000 engine Also, it may be benchmarking for similar gas turbine engines.

1 System description

The number of PW4000-94 produced has reached over 2500 since used in service in 1987. Nearly 2300 PW4000s were still in service in 2015. In this study PW4000 shown in Fig. 1 investigated has fan pressure ratio with 1.8 value, by-pass ratio with 5.1 value and total pressure ratio with 32 value during take-off. It produces about thrust at value of 260 kN at take-off [1]. The mass flow value of inlet air is 774 kg/s while, of this, 129 kg/s passes through the core engine, the remaining air is exhausted from fan exit. Air-fuel ratio is obtained as about 48.679.

* Corresponding author: onderturan@eskisehir.edu.tr

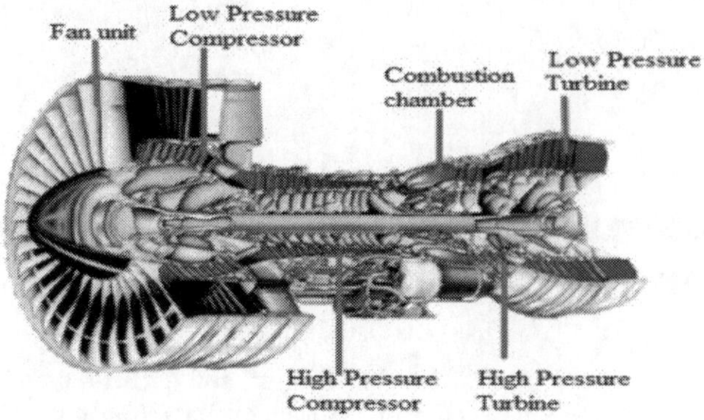

Figure 1. Illustration of PW4000 turbofan engine components [1]

2 Exergo-economic analysis of PW4000 engine

The exergoeconomic analysis arises out of combination of economic principles with second law of thermodynamics [2]. Both analyses together give more certain knowledge about system dealt with. The foundation of exergoeconomics analysis is that monetary costs associate with thermodynamic inefficiencies resulted from interactions emerged between system and its surroundings [3]. Namely, exergoeconomic methodology is employed to determine the costs of incoming and leaving streams, and exergy destructions within the system components [2]. The goal of this method is to optimize exergo-economic values for minimizing the exergy cost. Thus, the whole system has higher cost effectiveness. Exergo-economic behavior of the system can be evaluated for cost balance. Main equations related cost balance of an engine may be established as given below [4]:

$$\dot{C}_q + \int_i \dot{C}_k + \dot{Z}_k^T = \int_e \dot{C}_k + \dot{C}_w \tag{1}$$

$$\dot{C}_q = c\dot{E}_{q,k} \tag{2}$$

$$\dot{C}_k = c\dot{E}_{x,k} \tag{3}$$

$$\dot{C}_w = c\dot{W}_k \tag{4}$$

$$\dot{Z}_k^T = \dot{Z}_k^{CIC} + \dot{Z}_k^{OM} \tag{5}$$

Where $\dot{C}_q, \dot{C}_k$ and $\dot{C}_w$ represent the levelized exergy cost rate of heat transfer, entering or leaving stream and work done respectively. $\dot{E}_x, \dot{E}_q$ and $\dot{W}$ are the exergy of stream, heat transfer and power at inlet and outlet of the kth component. $\dot{Z}$ represents the levelised cost values for non exergetic with the units of $h-1. where T denotes total of capital investment cost (CIC) and operating & maintenance cost (OM) for the kth component.

3 Exergy assessment of PW4000 engine

The effiencieny in the energy use is very significant because all of the energy cannot be converted into useful work. Application of existing techonologies and several optimisation techniques together could provide benefit to reduce energy consumption [5]. The rise in energy efficiency means both the protection of environment from harmful emissions and economic saving for aviation sector. In fact, the limited resources lead to find out methods of sustainable development. For thermal systems, exergy evaluation is a foundation of economics, environmental and sustainability analyses, and thanks to results obtained, optimization of systems could performed on more factual base. Even only if applications of aircraft gas turbine engines are considered, the significance of an exergetic approach could be easily comprehended. Several exergy components exist in nature. The most used of them are physical and chemical exergies. Kinetic exergy is used only where velocity is high.

The meaning of physical exergy is the useful work provided from interaction between system and its surrounding, whereas chemical exergy could be defined as theorical work provided due to chemical composition difference between a system and its reference environment. Chemical exergy emerges only in combustion situations. If the first and the second laws of thermodynamics is applied on thermal systems, exergetic balance equality is easily established for related components. Physical exergy flow as called thermo-mechanical exergy divided into two components as thermal and pressure exergy. When ideal gas relations and constant specific heat capacity are accepted, this terms could be determined as follows respectively [6]:

$$E^{th} = \dot{m}c_p[(T_k - T_{amb}) - T_{amb} ln \frac{T_k}{T_{amb}}] \tag{6}$$

$$E^{pres} = \dot{m}RTln \frac{P_k}{P_{amb}} \tag{7}$$

In this study, when calculating thrust cost rate, exergies of Fan exit and exhaust is calculated with aid of kinetic exergy. The kinetic exergy of air and combustion gases is accounted for as follows [7]:

$$E^{pres} = \dot{m}RTln \frac{P_k}{P_{amb}} \tag{8}$$

Considering fossil fuels which is formulated as $C_a H_b$, their chemical exergy is defined as:

$$E^{ch} = \dot{m}_f \times LHV \times \left(1.033 + 0.0169 \frac{b}{a} - \frac{0.0698}{a} \right) \tag{9}$$

Whereas, chemical exergy of air constituents is found by [8]:

$$ex_{ch} = \sum_k x_k ex_k^{ch} + RT_0 \sum_k x_k ln x_k \tag{10}$$

In this context, exergy balance equilibrium for any control volume could be established as:

$$E^{ch} + \sum(E_o^{th} - E_i^{th}) + \sum(E_o^{pres} - E_i^{pres}) + \mathrm{T}_{amb} \sum(S_o - S_i) + \dot{Q} + \dot{W} = 0 \tag{11}$$

Exergy destruction required during efficiency calculations could be determined with aid of entropy production occured in components.

$$E_{xd} = T_{amb} \sum(S_o - S_i) \tag{12}$$

As for exergy efficiency, in literature various methods exist for this term. For in any component, this parameter could be defined by [8]:

$$\eta_{ex,k} = \frac{Ex_{pr}}{Ex_f} \tag{13}$$

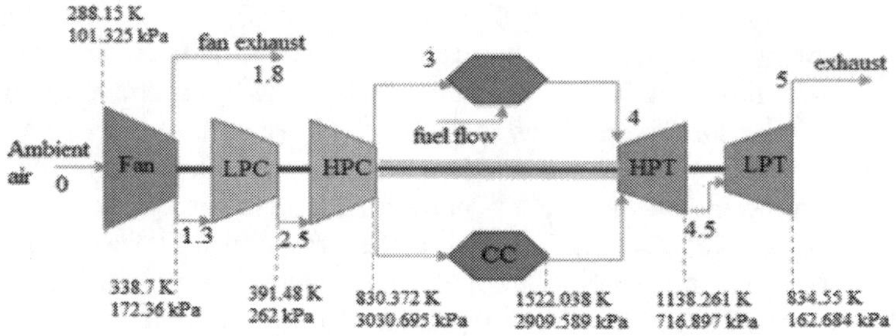

Figure 2. Cross-section drawing of PW4000 with thermodynamic informations

According to thermodynamics values given Figure 2, exergy analysis of PW4000 engine was carried out in advance [9,10].

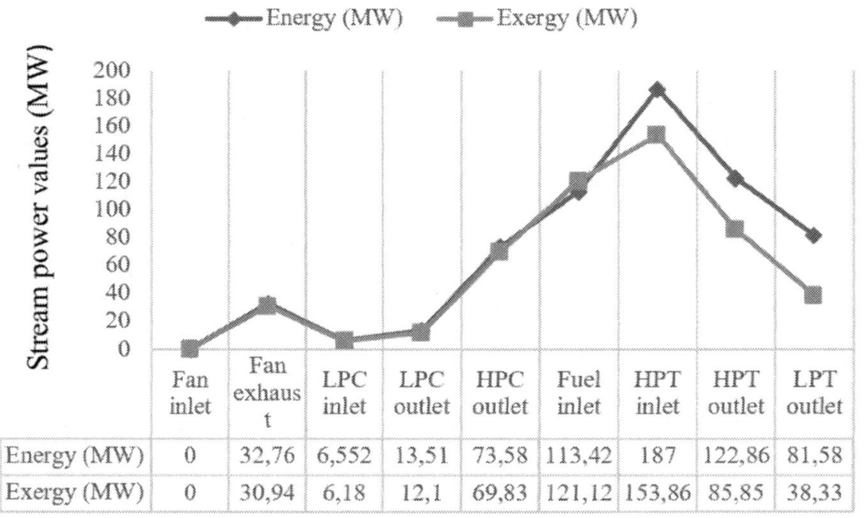

	Fan inlet	Fan exhaust	LPC inlet	LPC outlet	HPC outlet	Fuel inlet	HPT inlet	HPT outlet	LPT outlet
Energy (MW)	0	32,76	6,552	13,51	73,58	113,42	187	122,86	81,58
Exergy (MW)	0	30,94	6,18	12,1	69,83	121,12	153,86	85,85	38,33

Figure 3. Energetic and exergetic values of PW4000 engine for each station

4 Results and discussion

In this article, exergoeconomic analysis of high bypass PW4000 turbofan engine was carried out with aid of energy and exergy analysis performed in advance. The linear equation system solution and the relations between exergo-economic variables and economic performance parameters was provided developing code in the Matlab environment.

In Figure 4 cost rates is associated with the exergy destruction for each component. Thanks to two different methods, this parameter provide us opportunity to evaluate costs of these inefficiencies from different point views.

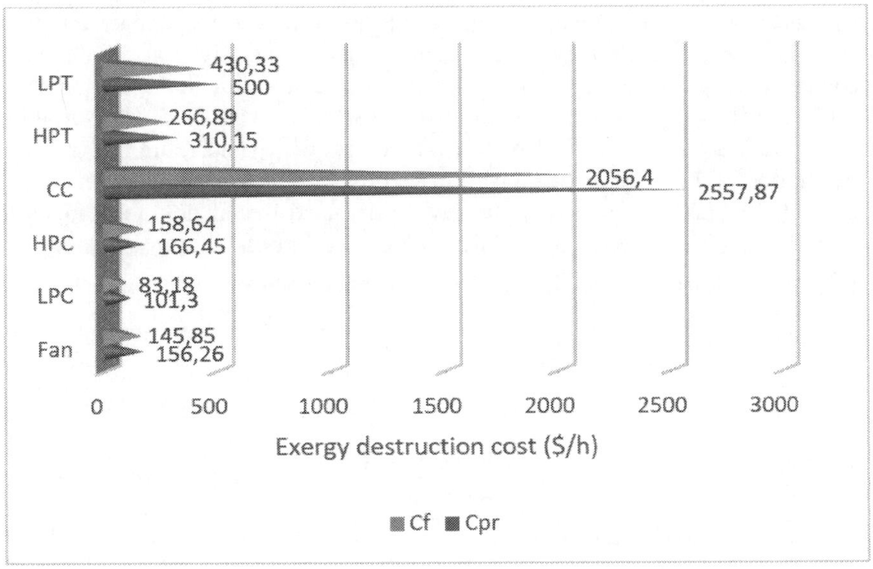

Figure 4. Exergy destruction cost rates of PW4000 engine components

Exergo-economic factor can display whether improvements to be performed is worth. This parameter is affected by two component. The first one is the exergetic costs resulted from inefficiencies. If this factor enjoys relatively low values, it should be known that exergetic costs outweigh. On the contrary, in high values of this factor, non-exergetic costs are dominant factor. According to Fig. 5, CC unit enjoys the lowest exergo-economic factor with 1.0801 %, while LPC unit has the highest ratio with 27.206 %. Fan, HPC and HPT stay modest values with 17.57 %,20.206 and 13.444 % respectively. Finally, LPT has second the lowest value with 6.738 %

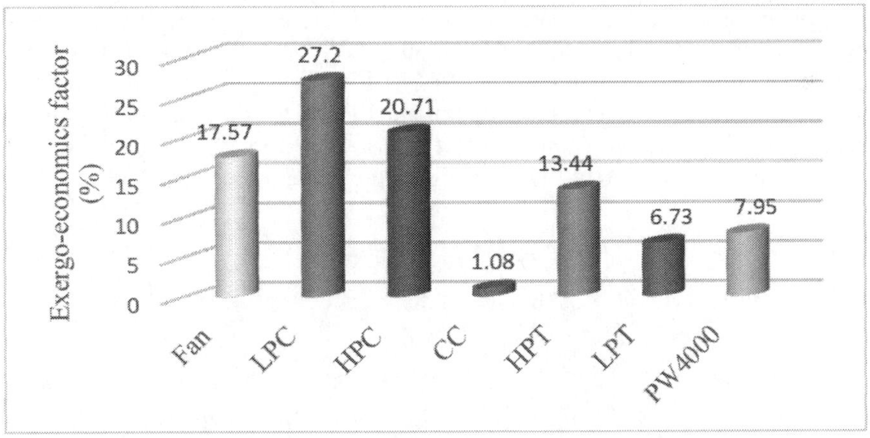

Figure 5. Exergo-economic factor values of PW4000 and its components (%)

For this engine, about three quarters of thrust is generated by Fan unit, whereas the rest amount of thrust is provided by combustion gases from exhaust. After find unit cost rate for Fan thrust with solution of linear equation system, corresponding unit cost for exhaust thrust can be calculated. Considering these calculations, cost rate for exhaust thrust is determined 4448 $/h, whereas for the Fan thrust, it is estimated as 1116 $/h. When these rates are divided to thrust values to find STC parameter, corresponding values are 68.45 $/hkN and 5.705

$/hkN respectively. Enormous difference in thrust cost rates could divert us to several initiatives such as increasing bypass ratio and fan pressure ratio to increase amount of fan thrust, but with current technology, some causes such as aerodynamic performance and sensitive balance between design variables do not allow us to achieve these attempt. As seen in Fig. 6, effect of the fuel price on specific thrust cost rate are handled. The variation in the range of 0.6 and 1.2 $/kg at the fuel price leads to rise the exhaust and fan specific thrust costs at the ratios of 110.239 % of 96.61% respectively. Corresponding rises are 82.4701 $/hkN and 5.4332 $/hkN respectively. It may be deduced that if new techniques towards increasing combustion efficiency or fuel derivations for turbofan engines are not explored, the aviation sector may not compete with the increase in fuel price.

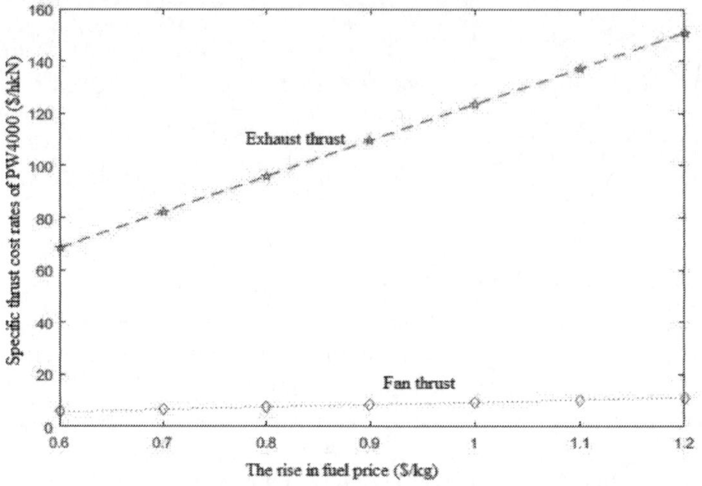

Figure 6. The variation in STC rates versus fuel price

Effect of PECs for HPC and HPT units on the fuel cost rates is displayed in Fig. 7. Although specific fuel cost ratios are not change much between 1.2 M$ and 2.4 M$, the variation in fuel cost ratios is calculated to be 80.12237 $/h and 75.03821 $/h respectively. These figures are highly crucial to consider PECs of components. Also these increases show parallelism with magnitude of exergo-economic factor which component has. Namely, HPC unit has relatively high exergo-economic factor. It is expected that it has higher rise in fuel cost rate according to variation in PEC. In Fig. 7, this situation emerged as mentioned.

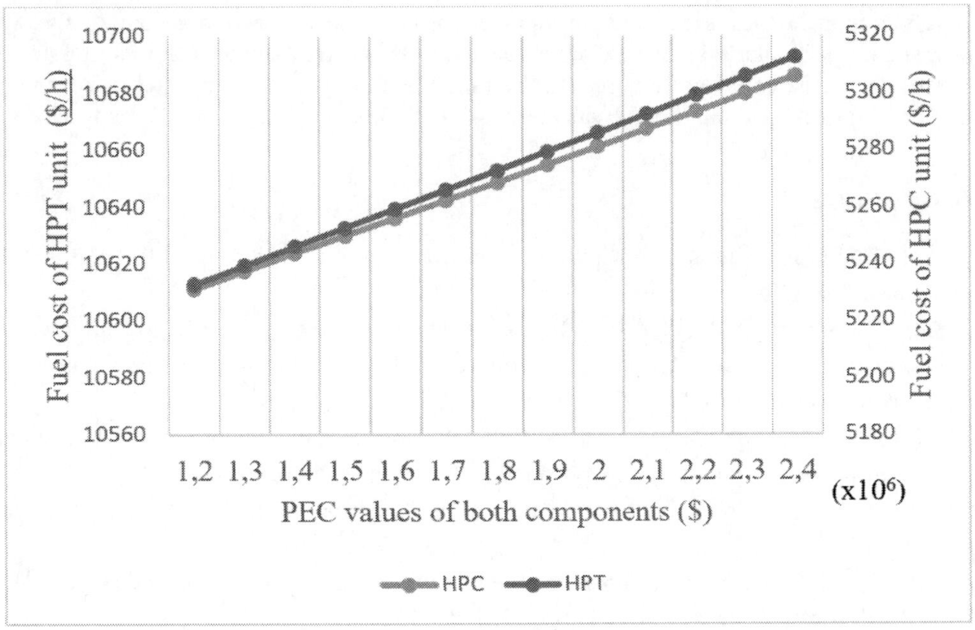

Figure 7. The change in fuel cost rates versus PEC value

5 Conclusions

The exergy analysis helps gain more point of view than the energy analysis about improvement of system. This analysis was performed in advance for PW4000 engine. If economic relations is added to this analysis, more meaningful and useful knowledge is obtained. Namely, whether the extra costs revealed from improvements could be sufficiently profitable could be determined.

• The highest value of exergy destruction cost rate as called exergoeconomic parameter belongs to CC with 2056.403 $/h according to fuel cost. Also LPC has the lowest value with 83.187 $/h.

• The relative cost ratio is important for CC unit because cost of outlet stream with 3.7507% is higher than inlet stream. Corresponding relative cost ratio is 24.386 %.

• One of the most crucial parameters among the others is exergo-economic parameter. The LPC unit is the highest magnitude with 27.206 %, while the lowest value belongs to CC unit with 1.0801 %.

• For HPC and HPT units, the variation between 1.2 M$ and 2.4 M$ in fuel cost rates versus purchased equipment cost investigated. The difference (increase) in fuel cost ratios with increasing PECs is found to be 80.12237 $/h and 75.03821$/h respectively.

• Impact of OM cost on exergo-economic factor and relative cost rate were discussed. The exergoeconomic factor of the LPC has the highest rise with 5.95 %, while the lowest variation value with 0.35 belongs to CC unit. As for relative cost rate, the variation in this parameter with the OM costs has the lowest values with 0.087 % for CC unit, while the LPC has relatively the highest value with 1.26 %.

• In the range of 3000 and 6200h, the diminish in exergoeconomic factors and relative cost rates are displayed. The highest the decrease emerges in LPC with 5.17 %, while the relatively lowest value belong to CC unit with 0.29 %.

• The magnitudes of STC parameter for exhaust and fan thrusts were determined to be 68.45 $/hkN and 5.705 $/hkN respectively.

• The variation in fuel price has a strongly effect on the exergoeconomic factor for all of components, particularly in the LPC with 16.9106%. Also effect of the fuel prices on the cost rates of the thrust could not ignore. In the range of 0.6 and 1.2 at the fuel price, the rise amount in exhaust and fan thrust costs are 82.4701 $/hkN and 5.4332 $/hkN respectively.

References

1. Pratt & Whitney [https://www.pw.utc.com/products-and-services/products/commercial-engines/PW4000-94-Engine/]
2. A.S. Almutairi , P. Pilidis , N. Al-Mutawa : Exergetic, Exergoeconomic and Exergoenvironmental Analysis of Intercooled Gas Turbine Engine. In: *52nd AIAA/SAE/ASEE Joint Propulsion Conference: 2016*; 2016: 5060.
3. S. Oyedepo , R. Fagbenle , S. Adefila : Exergetic and exergoeconomic analysis of selected gas turbine power plants in Nigeria. In*: 2014*: Citeseer; 2014.
4. A. Bejan, G. Tsatsaronis , M. Moran , MJ. Moran . Thermal design and optimization: John Wiley & Sons; 1996.
5. I. Dincer, M. Hussain, I. Al-Zaharnah. Energy and exergy use in public and private sector of Saudi Arabia. Energy Policy 2004, 32, 1615-1624.
6. S. Aliu, P. Ochornma. Exergoeconomic analysis of Ihovbor Gas Power plant. Nigerian Journal of Technology 2018, 37, 927-935.
7. O. Balli. Afterburning effect on the energetic and exergetic performance of an experimental turbojet engine (TJE). International Journal of Exergy 2014, 14, 212-243.
8. A. Bejan, G. Tsatsaronis, M. J. Moran. Thermal design and optimization; John Wiley & Sons, 1996
9. H. Aygun, O. Turan. Entropy, Energy and Exergy for Measuring PW4000 Turbofan Sustainability. International Journal of Turbo & Jet-Engines.
10. O. Balli . Exergy modeling for evaluating sustainability level of a high by-pass turbofan engine used on commercial aircrafts. Applied Thermal Engineering 2017, 123, 138-155.

MATEC Web of Conferences **314**, 02004 (2020) https://doi.org/10.1051/matecconf/202031402004
ICSC-ISATECH 2019

Finite element modelling of carbon fiber - carbon nanostructure - polymer hybrid composite structures

Stelios K. Georgantzinos[1,*], *Stylianos I.* Markolefas[1], *Stamatis A.* Mavrommatis[1], and *Konstantinos P.* Stamoulis[2]
[1]National and Kapodistrian University of Athens, General Department, Greece
[2]Amsterdam University of Applied Sciences, Faculty of Technology, Aviation Academy, Netherlands

Abstract. The present study deals with the numerical modelling of hybrid laminated composites, which can be proved especially useful in the engineering and maintenance of advanced aerospace primary structures. The lamina is comprised of continuous carbon fibers, thermosetting epoxy polymer matrix, as well as carbon nanostructures, such as graphene or carbon nanotubes, inclusions. Halpin-Tsai equations combined with results obtained from nanomechanical analysis are employed in order to evaluate the elastic properties of the carbon nanostructure/polymer matrix. Then, the obtained elastic properties of the hybrid matrix are used to calculate the orthotropic macro-mechanical properties of the unidirectional composite lamina. A hybrid composite plate is modelled as a 2D structure via the utilization of 4-node, quadrilateral, stress/displacement shell finite elements with reduced integration formulation. The convergence and analysis accuracy are tested. The mechanical performance of the hybrid composites is investigated by considering specific configurations and applying appropriate loading and boundary conditions. The results are compared with the corresponding ones found in the open literature, where it is possible.

1 Introduction

The high strength and stiffness, as well as the light weight of the polymer matrix composites make them ideal candidates in numerous automotive, aerospace, medical and other applications. Due to their excellent mechanical performance, carbon nanostructures, such as carbon nanotubes (CNTs) and graphene, can be used to reinforce the polymer matrix of composites, for a further enhancement of their stiffness and strength.

CNTs are cylinders formed by carbon atoms arranged in hexagonal pattern and being covalently bonded with each other through sp^2 hybridization. Their synthesis can be performed through several techniques. Arc evaporation method, laser ablation and chemical vapor deposition are the processes used broadly for production of CNTs [1]. CNTs present exceptional mechanical performance, i.e. remarkably high Young's modulus and strength, as well as low density [2,3].

* Corresponding author: sgeor@uoa.gr

Graphene has a similar carbon atom arrangement as CNTs; however, it is a planar monolayer. It is the thinnest structural material in the world and presents extraordinary mechanical [4,5], electrical [6], optical and chemical properties [7]. One can find various production and processing techniques, as well as potential applications of graphene in [8].

Nowadays, there has been an excessive growth in research associated with the polymer nanocomposites due to development of advanced materials for potential applications [9]. After establishing an increased understanding in the composition – structure – processing – performance dependencies for the class of the composite materials, a great effort has been devoted to enhancing the occasionally unsatisfactory matrix – dominate properties such as the impact damage [10-12]. Graphene or CNTs within the polymer matrix act as fillers which improves the engineering capabilities of a composite. Depending on the interaction between the CNT or graphene and the matrix, mechanical, thermal, magnetic and other properties of the nanocomposite improves substantially [13]. A recent review on polymer nanocomposites and their processing, characterization, and mechanical properties is presented in [14].

Finite Element Method (FEM) is one of the leading approaches for the simulation of physical phenomena and therefore can be straightforwardly used by both academic and industrial groups. Concerning the FEM simulation of the mechanical performance of polymer nanocomposites several works [2,15-17] have been presented in the literature, in which micromechanical analysis is performed. The advantage of the usage of FEM is its efficiency in terms of accuracy and computational effort.

Mehar and Panda [18] examined the deformation of CNT reinforced composite plate, numerically, experimentally and theoretically investigating the role of numerous parameters. They conclude that the bending rigidity increases by adding CNTs to the structure up to 0.2 weight fractions and decreases for higher ones. This phenomenon is explained to agglomeration of carbon nanotubes within the matrix. Hence, it is crucial to develop composite production methods, in which agglomeration does not occur. In an excellent work, Taş and Soykok [19] studied engineering constants of CNT - carbon fiber - polymer hybrid composite lamina and a composite plate configuration using ANSYS, ACP Module. The results have showed that engineering constants increase as the added carbon nanotube fraction is increased.

Beyond the extensive work about the investigation of carbon nanostructures or polymer nanocomposites, there are few studies concerning carbon fiber - carbon nanostructure - polymer hybrid composite structures. In this work, a FEM framework is presented for the analysis and design of hybrid composite laminated structures using conventional 4-node, quadrilateral, stress/displacement shell finite elements with reduced integration formulation. In order to describe the elastic constants of the hybrid composite, Halpin-Tsai equations combined with results obtained from nanomechanical analysis are used for the calculation of the carbon nanostructure/polymer matrix elastic properties and then the corresponding ones of the composite lamina. Assuming a specific plate configuration, the mechanical response of the composite is predicted under different loading conditions.

2 Modelling framework

The finite element modelling framework is based on the combination of the results obtained by mechanical analysis performed in different scales. Starting from the nanoscale, results concerning the elastic constants of CNTs and graphene are selected. Then, the nanoscale results are combined with the macroscale data concerning the polymer material using micromechanical rules, and specifically, here, the Halpin-Tsai model. In this way, the carbon nanostructure/polymer hybrid matrix material elastic constants are predicted. At that time, the mechanical properties of the unidirectional lamina of the hybrid composite can be calculated using the conventional macroscopic models. Therefore, the mechanical performance of a specific composite can predicted using the classical lamination theory and can be modelled using finite element techniques.

2.1 Calculation of elastic constants

2.1.1 Calculation of size dependent mechanical properties of nanostructures

The elastic mechanical behaviour of different sized carbon nanostructures as graphene and CNTs has been numerically investigated and predicted using a spring-based finite element approach [4, 20]. According to those approaches three-dimensional, two-node, spring-based finite elements of three degrees of freedom per node, are combined in order to model effectively the interatomic interactions presented inside a carbon nanostructure. The calculated variations of mechanical elastic constants have been approached by appropriate size-dependent, non-linear functions of two independent variables, i.e. length and width for graphene, in order to express the analytical rules governing the elastic behaviour. The numerical results have been already validated through comparisons with corresponding data given in the open literature. The significant effect of size and chirality of carbon nanostructures on their elastic behaviour has already been demonstrated.

The Young's modulus $E(w,l)$ in TPa of the graphene can be predicted using the following rational function

$$E_{gr}(w_{gr},l_{gr}) = \frac{P_0+a_{01}w_{gr}+b_{01}l_{gr}+b_{02}{l_{gr}}^2+c_{02}w_{gr}l_{gr}}{1+a_1w_{gr}+b_1l_{gr}+a_2{w_{gr}}^2+b_2{l_{gr}}^2+c_2w_{gr}l_{gr}},\ 0 < w_{gr}, l_{gr} \leq 10 \text{ nm}, \quad (1)$$

where $P_0, a_{01}, b_{01}, b_{02}, c_{02}, a_1, b_1, a_2, b_2,$ and c_2 are constants, which are determined by non-linear fitting, while w_{gr} and l_{gr} are the graphene width and length, respectively. The values of the constants of Equation (1) concerning both the two chirality directions of graphene are presented in Table 1. The accuracy of this equation compared with the finite element predictions and expressed by the coefficient of determination is calculated to be over 99.5%.

Table 1. Fitting parameters of Equation (1) for the prediction of Young's modulus variation in TPa.

Direction	P_0	a_{01}	b_{01}	b_{02}	c_{02}	a_1	b_1	a_2	b_2	c_2
Zig zag	676	13640	-2532	58250	19170	11760	-4.7E-4	63.72	85900	17870
Armchair	2.98	3.03	13.55	0.893	-1.197	3.042	19.53	1.148E-1	1.225	-1.634

2.1.2 Calculation of hybrid matrix elastic constants

Halpin-Tsai model proposed for discontinuous fiber reinforced lamina is employed to calculate the Young's modulus of the carbon nanotube/polymer matrix system. In this model, carbon nanotubes are considered as long randomly oriented discontinuous fibers. Young's modulus of the carbon nanotube-added matrix can be calculated from the following equation [19]:

$$E_{m-cnt} = \frac{3}{8}\frac{1+2\left(\frac{l_{nt}}{d_{NT}}\right)\left(\frac{\left(\frac{E_{eq}}{E_m}\right)-1}{\left(\frac{E_{eq}}{E_m}\right)+2\left(\frac{l_{nt}}{d_{NT}}\right)}\right)V_{nt}}{1-\left(\frac{\left(\frac{E_{eq}}{E_m}\right)-1}{\left(\frac{E_{eq}}{E_m}\right)+2\left(\frac{l_{nt}}{d_{NT}}\right)}\right)V_{nt}}E_m + \frac{5}{8}\frac{1+2\left(\frac{\left(\frac{E_{eq}}{E_m}\right)-1}{\left(\frac{E_{eq}}{E_m}\right)+2}\right)V_{nt}}{1-\left(\frac{\left(\frac{E_{eq}}{E_m}\right)-1}{\left(\frac{E_{eq}}{E_m}\right)+2}\right)V_{nt}}E_m, \quad (2)$$

where E_m is Young's modulus of matrix, V_{NT} is volume content of carbon nanotube and l_{NT} and d_{NT} are length and diameter of carbon nanotube, respectively. E_{eq} is equivalent modulus of the nanotube considering the hollow tube as a solid cylinder and can be expressed as

$$E_{eq} = \left(\frac{2t}{r}\right)E_{NT}, \quad (3)$$

where t, r, and E_{NT} are wall thickness, radius and Young's modulus of nanotube, respectively. Following equation developed by Tsai and Pagano can be used to calculate the shear modulus of the carbon nanotube-added matrix [19].

$$G_{m-cnt} = \frac{1}{8}\frac{1+2\left(\frac{l_{nt}}{d_{nt}}\right)\left(\frac{\left(\frac{E_{eq}}{E_m}\right)-1}{\left(\frac{E_{eq}}{E_m}\right)+2\left(\frac{l_{nt}}{d_{nt}}\right)}\right)V_{nt}}{1-\left(\frac{\left(\frac{E_{eq}}{E_m}\right)-1}{\left(\frac{E_{eq}}{E_m}\right)+2\left(\frac{l_{nt}}{d_{nt}}\right)}\right)V_{nt}}E_m + \frac{1}{4}\frac{1+2\left(\frac{\left(\frac{E_{eq}}{E_m}\right)-1}{\left(\frac{E_{eq}}{E_m}\right)+2}\right)V_{nt}}{1-\left(\frac{\left(\frac{E_{eq}}{E_m}\right)-1}{\left(\frac{E_{eq}}{E_m}\right)+2}\right)V_{nt}}E_m, \quad (4)$$

The Poisson's ratio for carbon nanotube-added matrix presenting exhibiting quasi-isotropic behaviour can be predicted from following equation

$$v_{m-cnt} = \frac{E_{m-cnt}}{2G_{m-cnt}} - 1. \quad (5)$$

Considering the mechanical response of the graphene/polymer matrix and assuming that the dimension of graphene inclusion is limited in nanoscale, the three dimensional (3D) Halpin-Tsai model for randomly oriented discontinuous rectangular reinforcements can be employed. According to this model, the Young's modulus of the graphene/polymer matrix can be predicted by the following equation

$$E_{m-gr(3D)} = \frac{1}{5}E_L + \frac{4}{5}E_T \quad (6)$$

where

$$E_L = \frac{1+\xi_L\eta_L V_{gr}}{1-\eta_L V_{gr}} E_m, \quad (7)$$

$$E_T = \frac{1+\xi_T\eta_T V_{gr}}{1-\eta_T V_{gr}} E_m, \quad (8)$$

and

$$\eta_L = \frac{\frac{E_{gr}}{E_m}-1}{\frac{E_{gr}}{E_m}+\xi_L}E_m, \ \eta_T = \frac{\frac{E_{gr}}{E_m}-1}{\frac{E_{gr}}{E_m}+\xi_T}E_m. \quad (9)$$

$$\xi_L = 2\frac{l_{gr}}{t_{gr}} + 40V_{gr}{}^{10}, \xi_T = 2\frac{w_{gr}}{t_{gr}} + 40V_{gr}{}^{10}, \quad (10)$$

while V_{gr} is the volume fraction of graphene in the polymer matrix.

Adopting the two-dimensional (2D) Halpin-Tsai model for randomly oriented discontinuous rectangular reinforcements, Young's and shear moduli can be calculated through the following equations

$$E_{m-gr(2D)} = \frac{3}{8}E_L + \frac{5}{8}E_T, \tag{11}$$

$$G_{m-gr(2D)} = \frac{1}{8}E_L + \frac{1}{4}E_T. \tag{12}$$

Assuming that Poisson's ratio for both 2D and 3D Halpin-Tsai models are approximately similar, it is calculated that

$$\upsilon_{m-gr} = \upsilon_{m-gr}(3D) = \frac{E_{m-gr(2D)}}{2G_{m-gr(2D)}} - 1. \tag{13}$$

Therefore, the shear modulus of the graphene/polymer matrix that exhibits quasi-isotropic behaviour can be calculated from the following equation

$$G_{m-gr(3D)} = \frac{E_{m-gr(3D)}}{2(\upsilon_{m-gr}+1)}. \tag{14}$$

2.1.3 Calculation of elastic constants of hybrid composite lamina

Young's modulus of a unidirectional lamina in longitudinal direction, E_1, and Poisson's ratio, υ_{12} can be calculated by utilizing the standard rule of mixture [21] as following

$$E_1 = E_f V_f + E_{m-cn}V_m, \tag{15}$$

$$\upsilon_{12} = \upsilon_f V_f + \upsilon_{m-cn}V_m, \tag{16}$$

where E_f, V_f, υ_f and V_m are Young's modulus of fiber, volume fraction of fiber, Poisson's ratio of fiber and volume fraction of matrix, respectively, while $E_{m-cn}= E_{m-cnt}$ or E_{m-gr} and $\upsilon_{m-cn}= \upsilon_{m-cnt}$ or υ_{m-gr}. The remaining orthotropic material properties, i.e. Young's modulus in transverse direction, E_2, Poisson's ratio, υ_{23} and shear moduli, G_{12} and G_{23} are calculated by the following Halpin-Tsai [22] equation

$$\frac{P}{P_m} = \frac{1+\xi\eta V_f}{1-\eta V_f}, \tag{17}$$

where P_m is the related properties of carbon nanostructure/polymer matrix. Moreover, η is an experimental factor, which is calculated by using following expression

$$\eta = \frac{\frac{P_f}{P_m}-1}{\frac{P_f}{P_m}+\xi}, \tag{18}$$

where P_f and P_m are the related properties of fiber and carbon nanostructure/polymer matrix, respectively. Note that ξ is a measure of reinforcement geometry which depends on loading conditions. ξ is chosen as 2 for calculation of E_2 and as 1 for calculations of υ_{23}, G_{12} and G_{23}.

2.2 Finite element modelling of a composite structure

Shell elements allow for modelling thin to moderately thick shells, down to a side-to-thickness ratio of 10. The Abaqus FEM program S4R conventional shell element is a 4-node, quadrilateral, stress/displacement shell element with reduced integration and a large-strain formulation. These elements are thick shell elements (non-zero transverse shear deformation) enforcing the first-order shear deformation theory (FSDT).

The first assumption of FSDT is that a straight line drawn through the thickness of the shell in the undeformed configuration may rotate but it will remain straight when the shell deforms. The second one is that the change of the shell thickness is negligible as the shell deforms. These assumptions are verified by experimental observation in most laminated shells when the ratio between the shortest surface dimension and the thickness, is larger than 10, as well as the stiffness of the laminae in shell material coordinates does not differ by more than two orders of magnitude.

Obtaining the elastic constants of the composite lamina and knowing the actual laminate stacking sequence (LSS) of a specific configuration, the laminate properties can be computed. A representative example of a finite element model is presented in Figure 1.

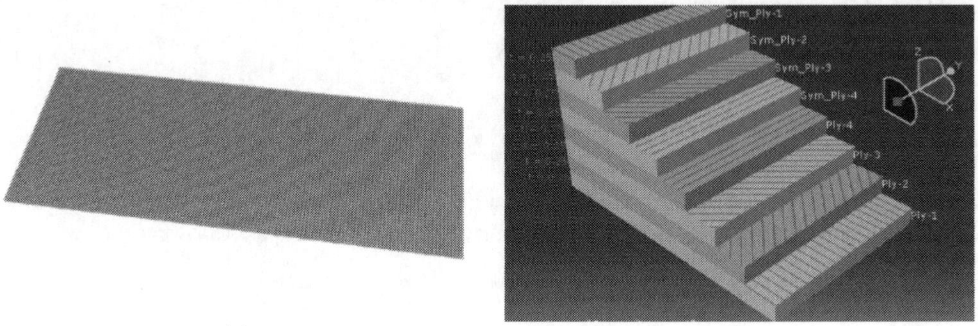

(a) (b)

Figure. 1. (a) Finite element model of a composite plate, and (b) LSS.

3 Results and discussion

In order to test the performance of the proposed method the following cases are considered. CNTs used in the model have l_{NT} = 20000 nm, d_{NT} = 20 nm, wall thickness t_{NT} =1.5 nm, and E_{NT} = 450000 MPa [23]. Graphene configuration and properties are described by w_{gr} = 10.02 nm, l_{gr} = 9.97 nm, and E_{gr} = 735000 MPa. Material properties of carbon fibers are E_{f11} = 230000 MPa, E_{f22} = 15410 MPa, v_{f23} = 0.46, v_{f12} = 0.29, G_{f12} = 10040 MPa, and G_{f23} = 10040 MPa [19]. Matrix material is assumed to be a polyester resin with E_m = 3000 MPa, G_m = 1139.8 MPa, and v_m = 0.316. Volume fraction of fiber and matrix is determined to be 60% and 40%, respectively.

Table 2. Material properties of the hybrid composite lamina with CNTs inclusions.

		CNT inclusion		Graphene inclusion	
V_{CN} (vol%)	0	0.2	0.5	0.2	0.5
E_1 (MPa)	139200	143610	150972	153000	187600
E_2 (MPa)	7799	14485	21450	23100	50250
G_{12} (MPa)	3194	7440	10514	11497	20560
G_{23} (MPa)	2560	5123	7254	8042	16449
υ_{12}	0.307	0.348	0.350	0.3073	0.307

		CNT inclusion		Graphene inclusion	
υ_{23}	0.404	0.450	0.452	0.4040	0.404

Material properties of the hybrid composite lamina with CNTs or graphene inclusions, calculated according to the proposed method are presented in Table 2. As expected, the hybrid composite with graphene inclusions exhibits better mechanical properties compared to CNTs inclusions because of the higher mechanical performance of graphene. Furthermore, graphene is a 2D structure and, thus, offers its advantages in two directions rather than the single direction of CNTs.

Table 3. Maximum bending deflection (mm) under different boundary conditions.

		CNT inclusion			Graphene inclusion
	Volume fraction	FEM (Present)	Theoretical [16]	FEM [16]	FEM (Present)
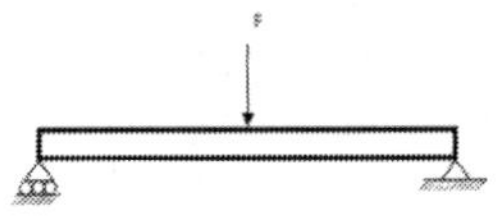	0	0.972	0.964	1.026	0.972
	0.2	0.865	0.860	0.918	0.774
	0.5	0.792	0.795	0.837	0.598
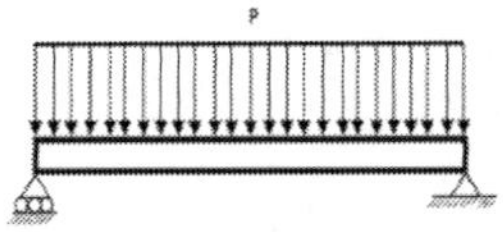	0	0.376	0.401	0.410	0.376
	0.2	0.345	0.362	0.367	0.324
	0.5	0.320	0.332	0.334	0.249

According to the ASTM D7264 standard, a 64 × 25 × 2 mm^3 composite plate is modelled. The plate is comprised of 8 unidirectional layers, possessing the mechanical properties computed in Table 2. The stacking sequence of the composite under consideration is [0/+45/-45/90]*s* (Figure 1b). The model has been tested for its convergence in terms of mesh density and the mesh depicted in Figure 1a is finally chosen. The hybrid composite plates of different carbon nanostructure inclusion types and volume fractions are subjected to bending loading ($F = 250$ N, $P = 0.1$ MPa). The analysis results obtained by the present method are compared with corresponding ones found in the literature in Table 3. A good agreement between the different methods is observed. To the author's best knowledge, numerical results concerning carbon fiber – graphene – polymer composites are presented for first time. Composite structures with graphene inclusions seem to present better mechanical performance than the ones with CNTs inclusions.

In Figure 2, the distribution of the bending deflection (for some cases of Table 3) are depicted.

Polyester resin ($V_{gr} = 0$)	Graphene/polyester resin matrix ($V_{gr} = 0.5$)
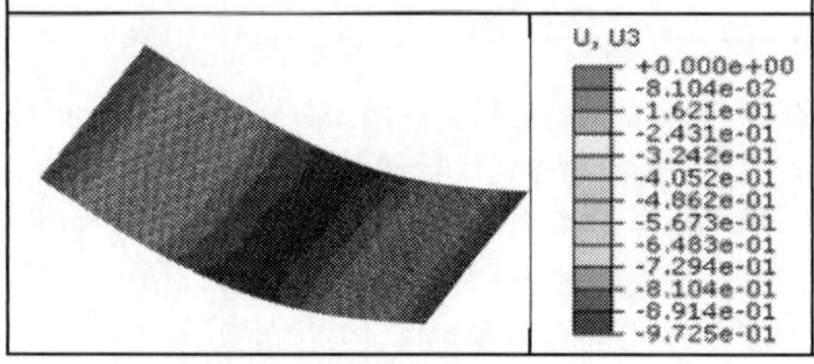	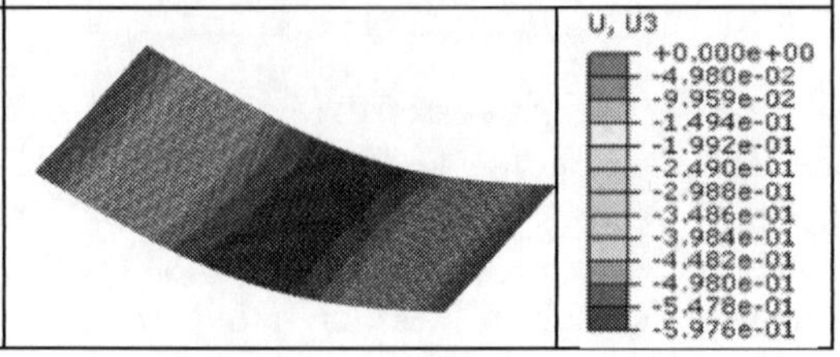

Figure. 2. Deformed mesh and deflection distribution of composite plates (F = 250 N).

4 Conclusions

A modelling framework for the simulation of the mechanical behaviour of carbon fiber - carbon nanostructure - polymer hybrid composite structures was presented. The proposed method can reveal the potential for the improvement of the mechanical performance of composite structures using carbon nanostructure inclusions. The stiffness of structures increases significantly due to the presence of nanostructures and graphene inclusion hybrid laminates demonstrate better mechanical performance compared to CNT hybrid laminates.

References

1. L.S. Ying, M.A.M. Salleh, H.B.M. Yusoff, S.B.A. Rashid, J.B.A. Razak, J. Ind. Eng. Chem. **17**, 367 (2013)
2. S.K. Georgantzinos, G.I. Giannopoulos, N.K. Anifantis, J. Comput. Theor. Nanos. **7**, 1436 (2010)
3. G.I. Giannopoulos, A.P. Tsiros, S.K. Georgantzinos, Mech. Adv. Mater. Struc. **20**, 730 (2013)
4. S.K. Georgantzinos, J. Comput. Theor. Nanos. **14**, 5347 (2017)
5. S.K. Georgantzinos, G.I. Giannopoulos, N.K. Anifantis, Proc. Inst. Mech. Eng. N J. Nanoeng. Nanosyst. **224**, 163 (2010)
6. S.K. Georgantzinos, G.I. Giannopoulos, A. Fatsis, N.V. Vlachakis, Physica E **84**, 27 (2016)
7. J.R. Potts, D.R. Dreyer, C.W. Bielawski, R.S. Ruoff, Polymer **52**, 5 (2011)
8. G. Mittal, V. Dhand, K.Y. Rhee, S.J. Park, W.R. Lee, J. Ind. Eng. Chem. **21**, 11 (2015)
9. P.M. Ajayan, O. Stephan, C. Colliex, D. Trauth, Science **265**, 1212 (1994)
10. K. Stamoulis, S.K. Georgantzinos, G.I. Giannopoulos, Int. J. Struct. Integr. https://doi.org/10.1108/IJSI-10-2018-0063 (2019)
11. 11. P. Dimoka, S. Psarras, C. Kostagiannakopoulou, V. Kostopoulos, Materials **12**, 1080 (2019)
12. 12. E. Giannaros, A. Kotzakolios, G. Sotiriadis, S. Tsantzalis, V. Kostopoulos. Compos. Struct. **204**, 745 (2018)
13. R.K. Layek, A.K. Nandi, Polymer **54**, 5087 (2013).
14. S. Fu, Z. Sun, P. Huang, Y. Li, N. Hu, Nano Mater. Sci. **1**, 2 (2019)
15. G.I. Giannopoulos, S.K. Georgantzinos, D.E. Katsareas, N.K. Anifantis, Comput. Model. Eng. Sci. **56**, 231 (2010)

16. A. Dabbagh, A. Rastgoo, F. Ebrahimi, Thin Wall Struct. **140**, 304 (2019)
17. G.I. Giannopoulos, S.K. Georgantzinos, P.A. Kakavas, N.K. Anifantis, Fuller. Nanotub. Car. N. **21**, 248 (2013)
18. K. Mehar, S.K. Panda, Adv. Polym. Tech. **37**, 1643 (2018)
19. H. Taş, I.F. Soykok, Compos. Part B-Eng. **159**, 44 (2019)
20. G.I. Giannopoulos, I.A. Liosatos, A.K. Moukanidis, Physica E **44**, 124 (2011)
21. R. M. Jones, *Mechanics of composite materials* (CRC press, 2014)
22. J.C. Halpin, J.L. Kardos, Polym. Eng. Sci. **16**, 344 (1976)
23. M.A. Rafiee, J. Rafiee, Z. Wang, H. Song, Z.Z. Yu, N. Koratkar, ACS Nano **3**, 3884 (2009)

1.2kW Interleaved synchronous buck converter design for aviation systems

Eralp Sener[1], *Isil* Yazar[2,*], *Gurhan* Ertasgin[1], and *Hasan* Yamık[1]
[1]Bilecik Seyh Edebali University, Department of Engineering, Turkey
[2]Eskisehir Osmangazi University, Department of Mechatronics, Turkey

Abstract. The principal aim of this work is to design a battery-operated 1.2kW interleaved synchronous buck converter for aviation systems to obtain lower losses compared to the existing topologies. It is desired to achieve at least a 5% loss reduction. Military aircrafts such as F-22 and civil aircraft as Boeing 787 are using the 270V DC for their battery systems. Although 270V DC system has some advantages such as reducing loses and the passive element sizes, old avionics that have been designed, produced and used since the second world war may not be compatible with the 270V DC system. Conversion from 270V to 28V appears to be quite promising for both old and new systems. Therefore, a buck converter design for 270 to 28V conversion is proposed in this study to adapt older avionics technology to accommodate new achievements. The analysis of the SiC MOSFET based interleaved synchronous buck converter with a double loop PI controller is presented. The paper demonstrates the principles of the design and discusses possible future developments. The concept is verified using simulations.

1 Introduction

The recent developments in aircraft applications are mainly replacing pneumatic and hydraulic applications with electrical equivalents [1-2]. 270V DC system which is recently employed in aircraft has many benefits. For instance, loss reduction in transmission and energy storage element sizing. It makes easier to supply the required power to the avionics with lower losses using 270V DC. It also makes easier to supply 115V AC to the system with smaller and less complicated inverters. However, the avionics have been designed, produced and used since the second world war may not be compatible with the 270V DC system. A buck converter can be used to adapt previous voltage levels and standards to enable the usage of those avionics. There are circuits in the market for the same voltage range but low power applications in general [3-6]. This reveals a requirement for circuit design with high power capability.

* Corresponding author: iyazar@ogu.edu.tr

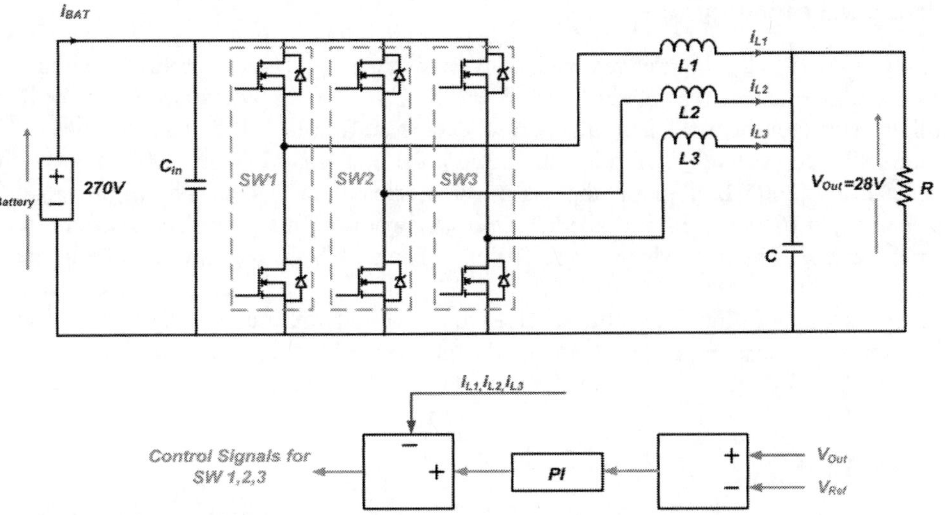

Figure. 1. The proposed interleaved synchronous buck converter

In this study, an interleaved synchronous buck converter topology is proposed for aircraft systems. Figure 1 shows this topology and its control block diagram.

1.1 Theoretical background

The weight and size of the power electronic elements can be quite critical while making a design for aircraft application. Interleaved systems cause smaller ripples at both input and the output of the circuit. So they use smaller energy storage elements compared to single buck converter's larger inductors or capacitors. They also provide higher reliability due to switch fault tolerance [7]. The output current is N times greater when each converter gives the same current as of the non-interleaved converter. The proposed interleaved topology is suitable due to the requirements for high current and low voltage output [7-9]. In addition, synchronous converter have many advantages compared to simple buck topology which has a diode instead of a high-frequency switch. Synchronous converters have higher efficiency compared to nonsynchronous converters with no forward voltage drop and they are better suited for high-frequency applications [10]. There are different topology approaches such as dual active bridge topology [11]. This topology is using wide-bandgap SiC semiconductors to increase the converter efficiency. Using GaN or SiC transistor switches for the design will result in substantial loss reduction in the converter system [12]. Therefore the designed topology employs SiC MOSFETs. Design requirements are defined by MIL-STD-704. The requirements for DC systems for both 28V DC and 270V DC are shown in Table 1. Transient voltage limits can be found in MIL-STD- 704 [13] specification.

Table 1. Voltage Standards [13].

	Limits	
	28 V DC System	**270 V DC System**
Steady State	22.0 To 29.0 Volts	250.0 To 280.0 Volts
Distortion Factor	0.035 Maximum	0.015 Maximum
Ripple Amplitude	1.5 Volts Maximum	6.0 Volts Maximum

1.2 Proposed topology

The study recommends 3 interleaved buck converter stages. All stages employ 2 switches for buck switching. Each switch couple (SW1, SW2, SW3) is driven by a double loop feedback controller. Each control signal acquires current data from IL1, IL2, IL3 and modulates SW1, SW2 and SW3 respectively. Utilizing the three-level interleaved stage, 100kHz switching frequency causes 300kHz ripples at the output waveforms of the system. This high-frequency ripple is easy to filter so it requires a smaller output capacitor. In addition, the output filter of the converter could be downsized if the driving signals of N converters have $2\pi/N$ phase-shift each [13]. The reason is due to including each converter's output current, and its frequency is greater than the non-interleaved converter frequency. Also, if a switch malfunctions, the remaining two stages of the interleaved system can continue operating within the MIL-STD-704 DC voltage standards [14].

2 Simulation

The interleaved synchronous buck converter is simulated using PSIM simulation software and is shown below in Figure 2. In the simulation, the switching losses are neglected. Circuit and simulation parameters are selected realistically to comply with MIL-STD-704 [14] standards. Ripple levels and voltage tolerances comply with the standards. Optimal working conditions are considered for element sizing and one switch fail scenario. In this paper, the optimal working scenario with the PI controller is simulated.

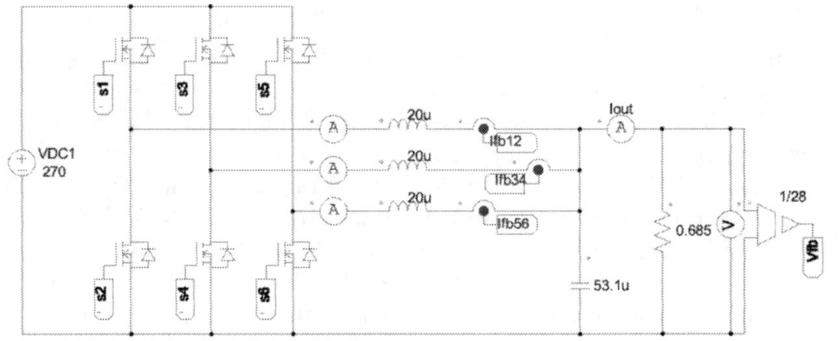

Figure. 2. The proposed circuit simulation model

The power stage of the circuit is shown in Figure 2. Inductors are selected for continuous conduction mode (CCM) with maximum 12A DC peak-to-peak ripple (ΔI) on each one of them. Also, an output capacitor is selected to have a maximum of 3% voltage ripple. 100kHz switching frequency is determined to use smaller passive components. As can be seen in Figure 2, the low-pass output filter consists of an inductor and a capacitor and provides decent filtering (see Figure 4). Discontinuous conduction mode (DCM) can also be employed with this topology using substantially smaller inductor value. This requires further investigation.

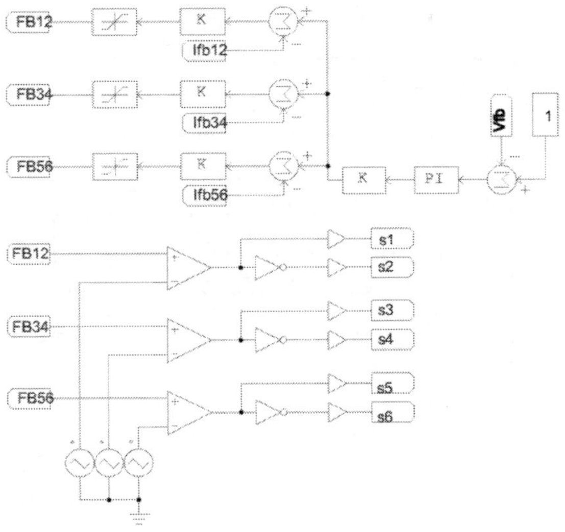

Figure. 3. Proposed circuit control and modulation scheme

Avionic systems require high-quality waveform and very stable voltage output. The converter literature suggests different methods for control systems. Voltage control applications are divided into two groups. One is the single loop voltage control [15-16] and the other is the double loop voltage control [17-18].

At the design stage of the control system, all interleaved systems considered as one system to protect the failure of the one from the other. Therefore, the control system uses three closed-loop PI control with current feedback as seen in Figure 3. Proportional-Integrator controller system is a control method which uses a reference signal to compare with the output using proportional and integrator calculations. In this controller, all K values are selected as 1, P is selected as 2 and I is selected as 0.004. Voltage feedback is divided by 28 and current feedback is divided by 100. If both voltage and current feedback loops are used, it is called double-loop control system. Voltage feedback is obtained from the output of the system which is directly connected to the load and current feedbacks for each interleaved stage is acquired directly from inductors. Positive inputs of the comparators are control signals of switches 1-2, 3-4 and 5-6 respectively. Negative inputs are 100kHz triangular carrier signals with $2\pi/3$ phase difference. It is important to note that if a failure occurs in a single interleaved stage, it is still possible to obtain the desired output voltage with this control approach.

3 Results

Figure 4 shows the converter output voltage (V_{out}) and current (I_{out}) waveforms. The topology and performed simulations comply with the standards at MIL-STD-704 [14]. System output value settles at the required value in 0.02 seconds. Also, 0.1A of current ripple (I_{ripple}) and 0.07V of voltage ripple (V_{ripple}) is obtained at the output. The conversion from 270V to 28V took place with an output current of about 41A and a power output of 1150W. Table 2 provides the parameters based on the simulations.

Table 2. Simulation Results.

Parameters	**Values**
V_{IN}	270V
I_{IN}	4.25A
V_{OUT}	28V
I_{OUT}	40.9A
V_{RIPPLE}	0.07V
I_{RIPPLE}	0.1A
Power	1145W

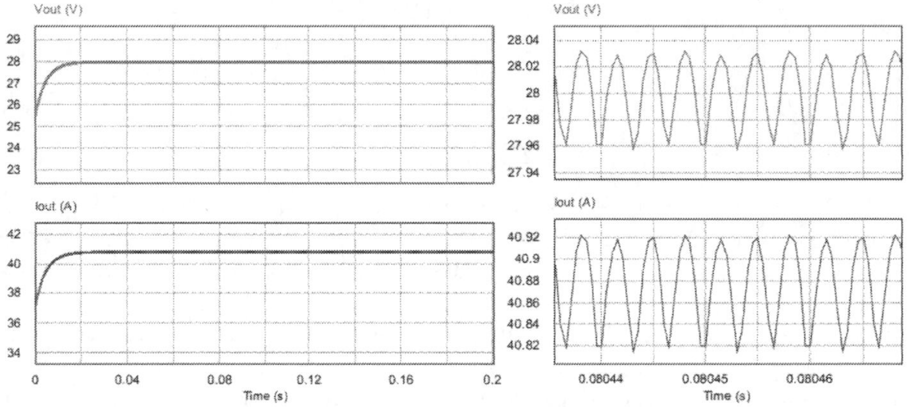

Figure. 4. Converter output voltage and current waveforms including ripples

Although the converter system demonstrates very high efficiency and almost no losses, all the passive elements and semiconductors are assumed to be ideal in the simulations. Future work requires loss calculations including thermal management and test results.

4 Conclusions

A 1.2kW synchronous interleaved buck converter designed and simulated for retrofit purposes. System design is based on reliability, better efficiency and standards which were determined by MIL-STD-704F [14].

A high-frequency switch is used for the low side switch instead of a diode to make the system interleaved. This permits optimal timing control and reduces the passive element sizes by allowing higher switching frequency so the losses can be reduced significantly. As a result, the proposed topology has reduced losses and fluctuations by employing synchronous and interleaved approaches whilst employing SiC MOSFETs. This also minimizes thermal management needs. Besides, it is still possible to obtain the required power output if there is a failure at one of the interleaved stages.

This topology and its variations for different power levels can be good candidates for the aviation industry. Although the proposed converter offers the desired loss reduction, further work is required to perform more detailed simulation and test results.

References

1. J. A. Rosero, J. A. Ortega, E. Aldabas and L. Romeral, "Moving towards a more electric aircraft," *IEEE Aerospace and Electronic Systems Magazine*, vol. 22, no. 3, pp. 3-9, Mar. 2007.
2. K. Emadi and M. Ehsani, "Aircraft power systems: technology, state of the art, and future trends," *IEEE Aerospace and Electronic Systems Magazine*, vol. 15, no. 1, pp. 28-32, Jan. 2000.
3. Vptpower.com. (2020). Retrieved 2 April 2020, from https://www.vptpower.com/vpt-products/cots-dc-dc-power-converters/vpthvm-270-bus-converter/.
4. DC-DC. Vicorpower.com. (2020). Retrieved 2 April 2020, from http://www.vicorpower.com/mil-cots/dc-dc.
5. MilCOTS DC-DC Power Converter | MCOTS-C-270-28-HT | SynQor. Synqor.com. (2020). Retrieved 2 April 2020, from https://www.synqor.com/products/mil-cots/mcots-c-270-28-ht.
6. Amcpower.com. (2020). Retrieved 2 April 2020, from http://www.amcpower.com/prod_dcdc_pbseries.html.
7. R. Giral, L. M. Salamero, S. Singer, "Interleaved Converters Operation Based on CMC," *IEEE Trans. Power Electron.*, vol.14, no. 4, pp.643-652, Jul. 1999.
8. M. T. Zhang, M. M. Jovanovic, F. C. Y. Lee, "Analysis and evaluation of interleaving techniques in forward converters," *IEEE Trans. Power Electron.*, vol.13, no. 4, pp. 690- 698, Jul. 1998.
9. J. Chen, D. Maksimovic and R. Ericson, "Analysis and Design of a LowStress Buck-Boost Converter in Universal-Input PFC Applications," *IEEE Trans. Power Electron.*, vol. 21, no. 2, pp. 320-329, Mar. 2006.
10. B. Sahu, and G. Rincon-Mora, "A Low Voltage, Dynamic, Noninverting, Synchronous Buck-Boost Converter for Portable Applications," *IEEE Trans. Power Electron.*, 19(2), 443-452.
11. D. De, A. Castellazzi, S. Lopez-Arevalo and A. Lamantia, "SiC MOSFET based avionic power supply," in *Proc. 7th IET International Conf. on Power Electron., Machines and Drives (PEMD 2014)*, Manchester, 2014, pp. 1-6.
12. R. R. Duarte, G. F. Ferreira, M. A. Dalla Costa and J. M. Alonso, "Performance investigation of silicon and gallium nitride transistors in an integrated double buck-boost LED driver," in *Proc. IEEE Industry Applicat. Society Annu. Meeting*, Cincinnati, OH, 2017, pp. 1-5.
13. M. Katayama, H Sekiya, and T. Yahagi, "An interleaved class E 2 dc/dc converter," in *Proc. IEEE International Symposium on Circuits and Systems*, pp. 2833- 2836, May. 2008.
14. MIL-STD-704F
15. X. Wang, P. C. Loh, F Blaabjerg. "Stability analysis and controller synthesis for single- loop voltage-controlled VSIs," *IEEE Trans. Power Electron.*, 2017, 32, (9), pp. 7394 - 7404.
16. X. Wang, Y.W. Li,.F Blaabjerg, et al., "Virtual-impedance - based control for voltage-source and current-source converters," *IEEE Trans. Power Electron.*, 2015, 30, (12), pp. 7019-7037.
17. X. Wang, F. Blaabjerg, Z Chen., "Resonance analysis in parallel voltage-controlled distributed generation inverters," *Proc. IEEE Appl. Power Electron. Conf. and Expo.*,

2013, pp. 2977-2983.

18. X. Wang, F. Blaabjerg, P.C. Loh, "High-performance feedback-type active damping of LCL-filtered voltage source converters," in *Proc. Energy Conversion Congress and Exposition*, Sep. 2015, pp. 2629-2636.

The assessment of commercial flights emissions at Ordu-Giresun Airport, Turkey

I. Orhan[1,*]
[1]Eskisehir Technical University, Faculty of Aeronautics and Astronautics, Turkey

Abstract. Ordu-Giresun International Airport is Turkey's first and world's second airport built on the sea. This study provides the pollutant gas emitted from aircraft with carbon monoxide (CO), hydrocarbons (HC), and nitrogen oxides (NOx) during landing and take-off cycles in Ordu-Giresun International Airport in 2015. It also presents the fuel burnt. For emission calculation, ICAO's database for aircraft engine emission data and General Directorate of State Airports Authority's database for flight records were used. In the calculations, a sensitivity analysis was made by taking into consideration the aircraft types and engine types used by the airline companies on the flights to Ordu-Giresun International Airport.

1 Introduction

Air transportation continues to grow with an increasing momentum from the past to the present. The passengers prefer as a priority transportation option with its fast mode of transportation between cities and countries. With the development of countries economic and welfare levels of people, the demand for tourist, commercial and cargo air transport is expected to continue to increase in the future. In 2017, 4.1 billion people made air travel in scheduled passenger transport worldwide. This figure means a 7.3% increase compared to 2016 [1]. According to IATA's 2017 report, in 2036, 7.8 billion passengers will travel using the airline [2].

In 2018, 12 million passengers are expected to carry out 120,000 flights daily. The $6 trillion goods transported in the airline industry in 2017 account for 35% of international trade. Globally, 57% of international travels are made by airline. In 2017, 341 billion litres of jet fuel, worth $ 149 billion, was used by commercial airlines. As a result, 859 million tons of carbon dioxide emissions were released in 2017 [3].

Kesgin [4] analysed exhaust emissions in the LTO (Landing and Take-off) phase that consists of four modes of engine operation, namely idle, approach, climb out and take-off in 40 Turkish national and international airport. In 2001, it was estimated that an average of 7614.34 to 8338.79 tons of exhaust emissions were produced annually during the LTO phase. In addition, it is stated that there is a 6% reduction in the exhaust emission of the LTO phase by reducing the taxi time by 2 minutes [4]. Tinmaz and etc [5] examined the exhaust emission values of the airplanes coming to Corlu Airport during the LTO phase. Yılmaz [6] made

* Corresponding author: iorhan@eskisehir.edu.tr

emission analysis in Kayseri Airport according to the data of the years of 2010. It was calculated that 8.4 t/y for HC, 66.90 t/y for CO, and 102.64 t/y for NOx were emitted during landing and take-off mode. Kuzu [7], estimated landing and take-off cycle emission at Atatürk International Airport and modelled the dispersion of emissions. The emissions of NOx, CO and HC were found as 4249, 2153, and 181t/y, respectively. It was shown that most of the airport areas and the residential area of the west of the airport exceeded the limit of NOx value annually. Postorino and etc [8] examined the taxi-out operation with stakeholders to reduce the emission of CO2 and propose some precautions to airport operation management.

In this study, the estimation of pollutant gas emissions of aircraft in LTO process in Ordu-Giresun Airport was investigated. ICAO - Engine emission database is used for emission calculations, and General Directorate of State Airports Authority's database for flight records in 2015 was used.

2 The calculation of aircraft emissions and air traffic at Ordu-Giresun airport

2.1 Ordu-Giresun Airport

Ordu-Giresun Airport is located in the Black Sea region of Turkey, close to the city center of Ordu and Giresun. The provinces of Ordu and Giresun are among the provinces that give a large amount of immigration both in Turkey and abroad. The majority of people working abroad and engaged in domestic migration are still in contact with their hometown. This is of great importance for the international passenger potential of the airport. Similarly, in Istanbul, Ankara, Bursa, Izmir and Kocaeli, a significant number of citizens live in Ordu and Giresun. Considering the fact that Ordu-Giresun Airport has a regional character, this situation is also crucial for the passenger potential of the airport. It is expected that Ordu and Giresun will become a preferred location for those interested in green and touristic areas and those interested in nature and extreme sports. The airport was opened in 2015 and has an annual capacity of 3 million passengers (Figure 1). Passenger and aircraft traffic information for the years between 2015 and 2019 is shown in Figure 2 and Figure 3 for Ordu-Giresun Airport.

Figure. 1. Ordu-Giresun Airport [9]

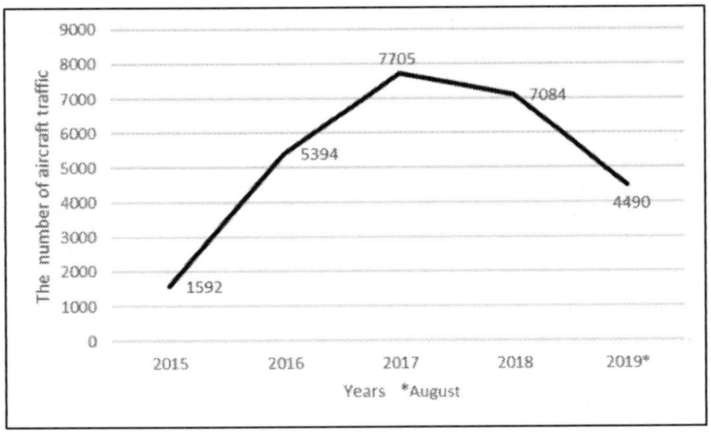

Figure. 2. The number of aircraft traffic at Ordu-Giresun Airport between 2015 and 2019 [10]

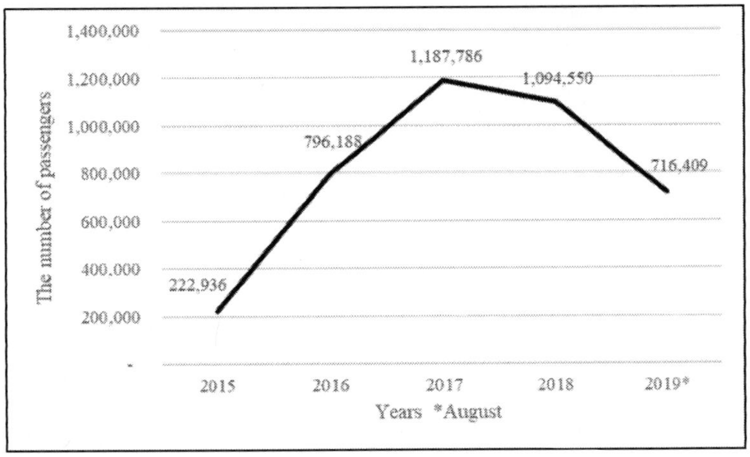

Figure. 3. The number of passenger at Ordu-Giresun Airport between 2015 and 2019. [10]

2.2 Methodology and Calculation

The amount of emissions emitted and the amount of fuel consumed by an airplane in the landing and take- off cycles varies depending on many variables such as weather conditions, traffic density at the airport and the types of aircraft using the airport and the time the aircraft spends for the required power in the cycle. The landing take-off cycle defined by ICAO is shown schematically in Figure 4.: Approach, taxi, take-off, climb.

According to the ICAO engine emission certification procedure, the LTO cycle comprises the following power setting and time-in mode [11].

- Take-off: 100% thrust for 0.7 minutes
- Climb: 85% thrust for 2.2 minute
- Approach: 30% thrust for 4 minutes
- Taxi: 7% thrust for 26 minutes. Taxi consists of two parts taxi-in and taxi-out

The approach consists of operations on the airport from 3000 ft until the planes reach the runway. The taxiing phases consists of the maneuvers of the aircraft up to the take-off position and those that are blocked until it reaches the runway threshold. Take-off includes the process from runway threshold to aircraft rotation. The climb includes from rotation until the plane reaches 3000 ft.

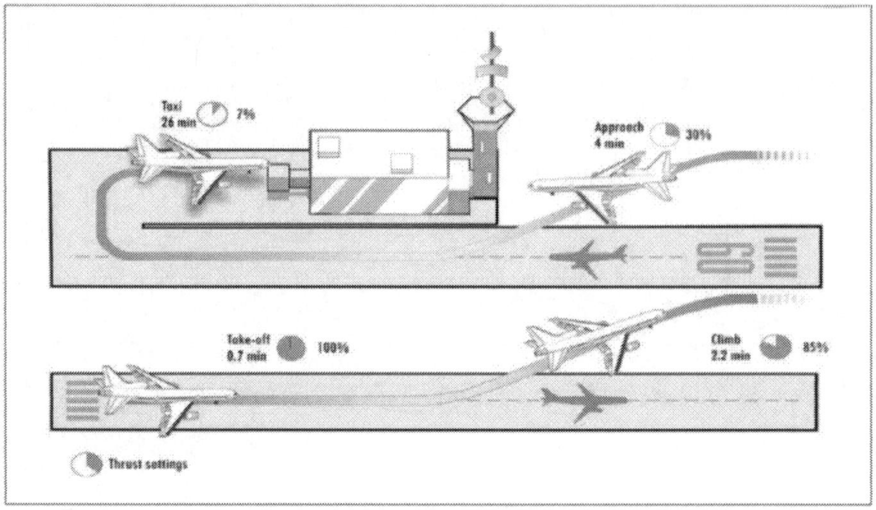

Figure. 4. The landing take-off cycle [12]

The fuel consumption and emission indices of an aircraft for each operating mode are taken from the ICAO-Engine emission database [13]. The equations (1) is used for emissions calculation (4,14,15);

$$E_{i,m} = \sum_a \sum_e n_a l_{a,e} F_{a,,} E_{e,m,i} t_{m,a} \quad (1)$$

Where $E_{(i,m)}$ is annual emission of pollutant i for mode m, (kg/yr); n_a is the number of engines of aircraft type a; $l_{(a,e)}$ is the number of annual LTO cycles for aircraft type a with engine type e; $F_{a,e,m}$ is fuel consumption for aircraft type with engine type e in mode m, (g/kg); $E_{(e,m,i)}$ is emission factor for engine type e and mode m and pollutant i, (g/kg) and $t_{(m,a)}$ is time in mode m for aircraft type a, (s).

Ordu-Giresun Airport was opened on May 22, 2015 and 1592 commercial flights operated in its first year. According to 2015 data, the aircraft used in the commercial flights to the airport and the engine types used in the aircraft is shown in Table 1.

Table 1. The aircraft with engine type used for flight Ordu-Giresun Airport

A320-200	CFMI CFM56-5B4/3
B737-800	CFMI CFM56-7B24
B737-800	CFM56-7B27
B737-800	CFM56-7B26
B737-800	CFM56-7B26E

3 Results and conclusion

In this study, the pollutant emissions and the amount of fuel consumed during the LTO cycle at Ordu-Giresun Airport in 2015 are considered. The mass of fuel burnt, the mass of CO, HC and NOx that emitted were calculated and presented in Table 2.

Table 2. The aircraft with engine type used for flight Ordu-Giresun Airport

Mass of fuel burnt (kg)	1,387,485.31
Mass of CO emitted (kg)	14,375.86
Mass of HC emitted (kg)	1,083.52
Mass of NOX emitted (kg)	17,537.58

The demand for air transport increases with an annual average increase of 5% around the world. The increasing number of flights due to passenger demand will be one of the significant causes of air pollution in the airport area. At the airport, emissions analysis from airplanes was conducted in the year the airport was started to operate, with relatively low air traffic. Accordingly, because of the lack of heavy traffic, the taxi duration of the aircraft will also be less than ICAO's time-in-mode values, thus reducing the emission rates in part. Emissions emitted from aircraft will increase due to increased taxi time at the airports, stop and go flights, and the amount of power required for the return movements. Providing faster and more direct taxi routes with fewer stops may be the first step to reduce emissions at the airport. Compared to the previous years, although the use of new aircraft and new engines, emissions are still lagging behind the targets, partly due to the fact that old technology engines are still in use. With the development of combustion and engine technology and aircraft designs, efforts are underway to further reduce emissions in the future. However, minimizing potential delays at airports will also contribute to reducing emissions.

References

1. Rosen, E. Over 4 Billion Passengers Flew In 2017 Setting New Travel Record. (2018) https://www.forbes.com/sites/ericrosen/2018/09/08/over-4-billion-passengers-flew-in-2017-setting-new-travel-record/#235f8c2a255b
2. IATA. 2036 Forecast Reveals Air Passengers Will Nearly Double to 7.8 Billion. (2017). https://www.iata.org/pressroom/pr/Pages/2017-10-24-01.aspx
3. ATAG. Aviation Benefits beyond Borders. (2018) https://aviationbenefits.org/media/166344/abbb18_full-report_web.pdf
4. Kesgin U., Aircraft emissions at Turkish airports, Energy, **31**, 372–384, (2006)
5. Tınmaz E, Özkan A, Akpınar A., Aircraft originated air pollution in the example of Çorlu airport, Journal of Environmental Protection and Ecology, **3** (3), 586-592, (2002)
6. Yılmaz, İlker., Emissions from passenger aircraft at Kayseri Airport, Turkey, Journal of Air Transport Management, **58,** 176-182, (2017)
7. Kuzu, S. Levent., Estimation and dispersion modeling of landing and take-off (LTO) cycle emissions from Atatürk International Airport, Air Quality, Atmosphere & Health, **11**(2), 153-161, (2018)
8. Postorino, Maria Nadia, Luca Mantecchini, and Filippo Paganelli., Improving taxi-out operations at city airports to reduce CO2 emissions, Transport Policy **80**, 167-176, (2019)
9. https://www.airporthaber.com/ordugiresun-havalimani-haberleri/, (2019)
10. https://www.dhmi.gov.tr/sayfalar/istatistik.aspx, (2019)
11. International Civil Aviation Organization, Exhaust emission data bank, Doc. 9646-AN/943, Montreal, CA, (1995). See also www.icao.org.

12. ICAO, Environmental Report Aviation And Environment, (2019) https://www.icao.int/environmental-protection/Documents/ICAO-ENV-Report2019-F1-WEB%20%281%29.pdf
13. ICAO Aircraft Engine Emissions Databank, (2019) https://www.easa.europa.eu/easa-and-you/environment/icao-aircraft-engine-emissions-databank
14. Stefanou, P., and D. Haralambopoulos, Energy demand and environmental pressures due to the operation of Olympic Airways in Greece, Energy, **23**.2, 125-136, (1998)
15. Ekici, S., Yalin, G., Altuntas, O., Karakoc, T. H., Calculation of HC, CO and NOx from civil aviation in Turkey in 2012, International Journal of Environment and Pollution, **53** (3-4), 232-244, (2013)

An intelligent data filtering and fault detection method for gas turbine engines

Amare D. Fentaye[1,*], and *Konstantinos G.* Kyprianidis[1]
[1]School of Business, Society and Engineering, Mälardalen University, 883, SE-72123 Västerås, Sweden

Abstract. In a gas turbine fault diagnostics, the removal of measurement noise and data outliers prior to the fault analysis is very essential. The conventional filtering methods, particularly the linear ones, are not sufficiently accurate, which might possibly lead to the loss of critically important features in the fault analysis process. Conversely, the recorded accuracies obtained from the non-linear filters are promising. Recently, the focus has been shifted to the artificial neural network (ANN) based nonlinear filters due to their capability of providing a robust identity map between the input and output data, which can be efficiently exploited in the process of fault diagnosis. This paper aims to present combined auto-associative neural network (AANN) and K-nearest neighbor (KNN) based noise reduction and fault detection method for a gas turbine engine application. The performance of the developed method has been evaluated using data obtained from a model simulation. The test results revealed that the developed hybrid method is more effective and reliable than the conventional methods for the fault detection of the gas turbine engine with negligible false alarms and missed detections.

1 Introduction

Performance deterioration of a gas turbine engine strongly affects its operation. Gas path components are subjected to a very harsh operating conditions and environmental contaminants that may significantly degrade their performance [1]. Gas path component's performance deterioration can be expressed in terms of performance parameter (such as compressor flow capacity, compressor isentropic efficiency, turbine flow capacity and turbine isentropic efficiency) changes [2]. These changes would result in changes in gas path measurements (temperature, pressure, fuel flow, and shaft speed). Degradations are manifested by trend changes in gas path measurements from their baselines. Thus, in order to maintain a reliable engine operation, the abnormal operating conditions has to be detected successfully. However, as measurement uncertainties influence the detection accuracy, an effective data filtration prior to the fault detection task is also crucial.

Kalman filter (KF) is one of the conventional model based approaches used to undertake noise reduction and data validation tasks. For instance, multiple KF models were used for

* Corresponding author: amare.desalegn.fentaye@mdh.se

sensor and actuator fault detection and isolation (FDI) together with a component fault detection in an aircraft engine by Kobayashi and Simon [3]. A single FDI effectiveness of KF, artificial neural network (ANN) and hybrid ANN was investigated by Volponi et al. [4]. Although three of them could isolate more than 90% of the fault signatures, the result from the hybrid system was the best. Pourbabaee et al. [5] also proposed a multi-model hybrid KF based engine sensor fault detection, isolation and identification (FDII) method. However, the effectiveness of the KF based techniques, particularly the linear filters, were highly influenced by the nonlinearity of the data and the level of measurement uncertainties [6]. To address the issue of nonlinearity, several authors like Lu et al. [7, 8] utilized the modified version of the linear KF. On the other hand, regardless of its limitation to deal with measurement uncertainty, gas path analysis (GPA) has also been widely utilized for engine performance diagnostics [9-12]. As far as artificial intelligence (AI) methods are concerned, Kramer [13, 14] introduced an AANN based technique for sensor validation and correction that is coping with the nonlinearity of the data. He used the network residuals to detect and estimate sensor faults. In the same year, Lu et al. [15] implemented an AANN for a gas turbine engine data screening and validation, and promising results were reported. Ganguli [16] also employed Median filters (MFs) and Fuzzy Logic (FL) for engine data filtering and fault detection. The MF scheme was used to filter measurement noises and outliers, while the FL based method used to detect engine component faults based on gas path measurement residuals. Recently, more attention has been given to ANNs due to their performance in noise reduction and FDI [17-22]. On the other hand, application of KNNs for a gas turbine diagnostic is also started lately. About a decade ago, the potential of KNN classifier for a two-spool turbofan engine fault classification was evaluated and compared with support vector machine (SVM), principal component analysis (PCA), and probabilistic neural network (PNN) methods by Donat et al. [23]. On another study, Loboda [24] compared the gas turbine fault classification performance of KNN, multilayer perceptron (MLP), and PNN. In general, combining two or more methods showed different advantages in gas turbine diagnostics than using the methods individually [25, 26].

In this paper, a combined AANN and KNN method is utilized for a gas turbine noise reduction and fault detection. The effectiveness of the method has been tested by applying on a two-shaft stationary gas turbine engine, which is thermodynamically similar to LM2500. The required training and test data samples are generated by simulating the performance model of the case engine under steady state operating conditions. In order to take into account measurement noises, Gaussian noise is added to the data associated with each measurement parameter.

2 Method

The proposed noise reduction and fault detection technique as shown in Figure 1 is developed by combining AANN and KNN approaches. The AANN structure is used to minimize the measurement noise and the KNN structure to distinguish the fault and no-fault data classes. Using the AANN structure prior to the fault detection may help to enhance the detection effectiveness. Because one of the major sources of a wrong detection is the occurrence of a high-level sensor noise due to the harsh engine operating environment. For an effective and practically acceptable fault detection, the noise effect should thus be minimized. During real-time application, first, the observed measurements will pass through the AANN structure for de-noising, next the residuals will be generated by taking the difference between the AANN output and the established baseline, then lastly those residuals will be fed into the classifier to decide whether the engine is healthy or unhealthy.

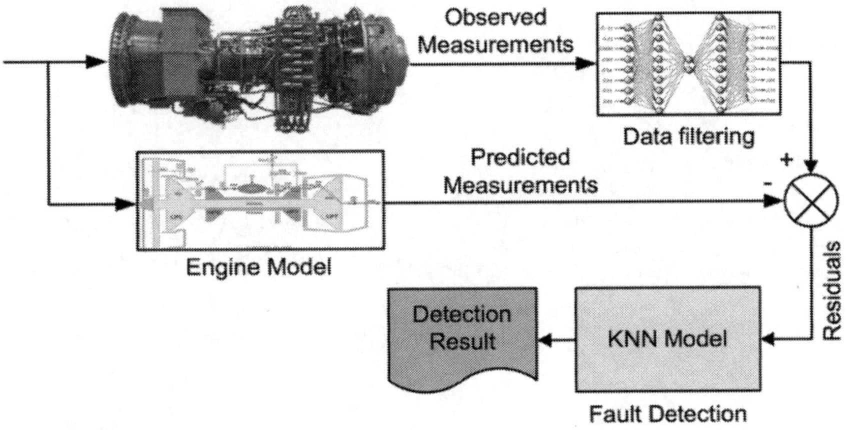

Figure. 1. Schematic diagram of the noise filtering and fault detection scheme.

2.1 AANN-based noise reduction

AANNs are feed forward types of neural networks (NN) that try to learn an approximation to the identity function in an unsupervised manner so as to give an output that is similar to its input [14]. Figure 2 shows the general topology of a five-layer AANN. The overall structure has two parts, compression and decompression. During the forward propagation, the compression part takes the high dimensional input information and compresses it into a smaller dimension with capturing important features. The decompression part then takes the compressed data from the bottleneck layer and tries to reconstruct the original information. It composed of an input layer, a mapping layer, a bottleneck layer, a de-mapping layer, and an output layer. The output layer is of the same size as the input layer. The bottleneck layer is located in the middle of the network, with the smallest number of neurons, where the important features of the data captured. The number of neurons in the bottleneck layer is determined based on the principal components required to extract the sufficiently important features of the input data. The sizes of mapping and de-mapping layers depend on the complexity of the network functions that can be generated during learning. According to the universal approximation theorem, the activation function for the mapping and de-mapping layers should be non-linear, whereas the remaining layers can use linear activation function. More details on AANN can be found in [13].

During training, noise-corrupted data is used as an input and noise-free data as an output. Using noisy data on both sides reduces the effectiveness of the de-noising network [27]. The network is then tried to reduce the noise and provide an output, which is as close as the desired noise-free output.

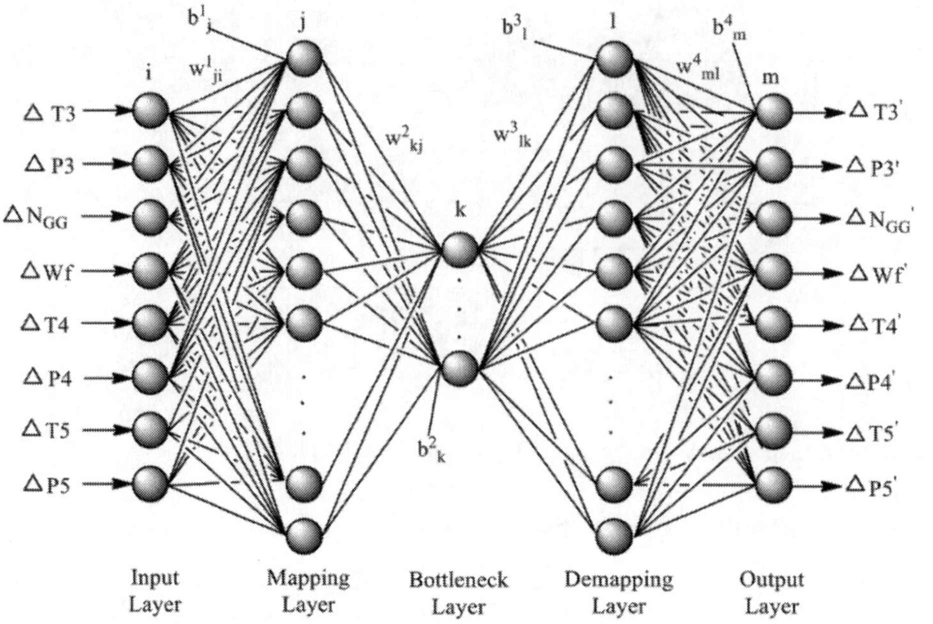

Figure. 2. The general five layer AANN architecture with 8 input/output parameters.

2.2 Engi ne fault detection using KNN

Fault detection is the first and the very important step in fault diagnostics that used to decide whether the operating system is healthy or faulty. A detection system with a minimum false alarm and missed detection is thus very essential. In the current work, the fault detection problem is considered as a binary class classification problem. A KNN based classification technique is utilized to distinguish the healthy and faulty data classes of the engine. The process of developing a KNN classifier requires two steps: building the model using the training dataset and testing its performance on the test dataset. New data points are predicted based on their distance from the neighbor classes. Figures 3 and 4 show the conceptual model of a 2-class KNN classifier with different K values and the complete procedure used to develop the model, respectively.

The brief description of the major steps involved in the process of developing the classifier is given as follows:

Step 1. The feature vectors associated with the healthy and faulty classes are prepared. The whole data is then divided into two subsets, one to construct the model and the rest to test its classification accuracy. The class labels are defined for the training dataset and saved in the database together with the unlabeled test feature vector.

Step 2. The model is constructed based on the known features and labels of the training dataset to visualize the two classes.

Step 3. The unlabeled test data samples are used to evaluate the classification performance of the model. In order to determine the class where the test data belongs to, the distances between the new data points and all the training data points need to be computed. Euclidian, Minkowski, Manhattan, Cosine distance, and Correlation distance are the most commonly used distance functions for KNN.

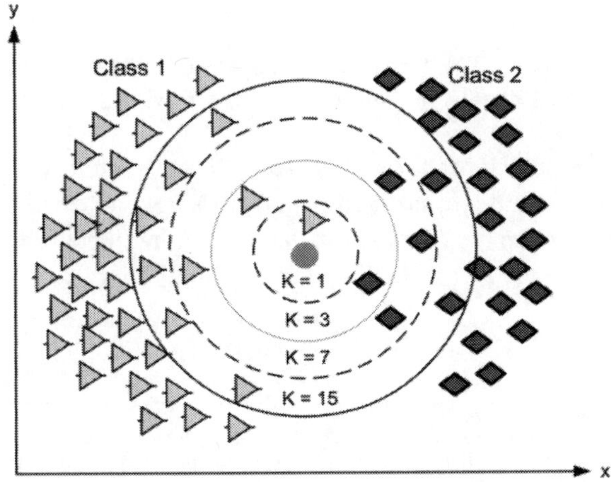

Figure. 3. Conceptual model of KNN based classification.

Step 4. The computed distances corresponding to each class were sorted in an ascending order, and the first k nearest neighbors were selected based on their distance. When k is 1, the class of the new data point will be determined based on the first nearest point. It is preferable when a very large training data is used. Too big k values may result over generalization, while too small k values usually cause overfitting and unstable decision boundaries. k should always be an odd number to avoid tie in the vote.

Step 5. After selecting the appropriate k nearest neighbors, the class of the test data can be decided according to the majority vote, by counting the number of nearest neighbors belonging to each class or weighted-sum voting. If weighted distances are used to make the vote, the degree of contribution of the nearest neighbors should be computed applying different distance weighting functions. The weight value of each neighbor is added to the total weight of the corresponding class and the class with the largest weight value wins. One of the following distance functions can be used to weight the contribution of neighbors: equal (weight = 1), inverse (weight = 1/d), and square inverse (weight = 1/d2).

Step. 6. In order to select the most appropriate classification model, the cross-validation technique was applied. There are different cross-validation techniques available for machine learning algorithms. Two of the most widely used techniques are holdout cross-validation and k-fold cross-validation. The holdout cross-validation is accomplished by dividing the sample dataset into two sub-sets, training and test datasets. Whereas, in the latter case, the sample data is divided into k number of equal-sized groups or folds. The training is done on the k-1 folds while remaining fold is left for validation. The latter approach was applied in the current work.

Step 7. The detection accuracy was then evaluated based on the following Equations:

$$RTP = \frac{TPs}{TPs + FNs} \times 100\% \tag{1}$$

$$RFP = \frac{FPs}{FPs + TNs} \times 100\% \tag{2}$$

$$Missed\ Detection = [100 - RTP]\% \tag{3}$$

$$Faulse\ Alarm = [100 - RFP]\% \tag{4}$$

$$ODA = \frac{TPs + TNs}{TPs + FPs + FNs + TNs} \times 100\% \tag{5}$$

where:

- RTP = rate of true positive, RTN = rate of true negative, RFP = rate of false positive, RFN = rate of false negative, and ODA = overall detection accuracy.
- TPs (True positives) = number of correctly detected faulty patterns.
- FPs (False positives) = number of no-fault patterns wrongly detected as faulty.
- TNs (True Negatives) = number of correctly detected no-fault patterns.
- FNs (False Negatives) = number of fault patterns wrongly detected as no-fault.

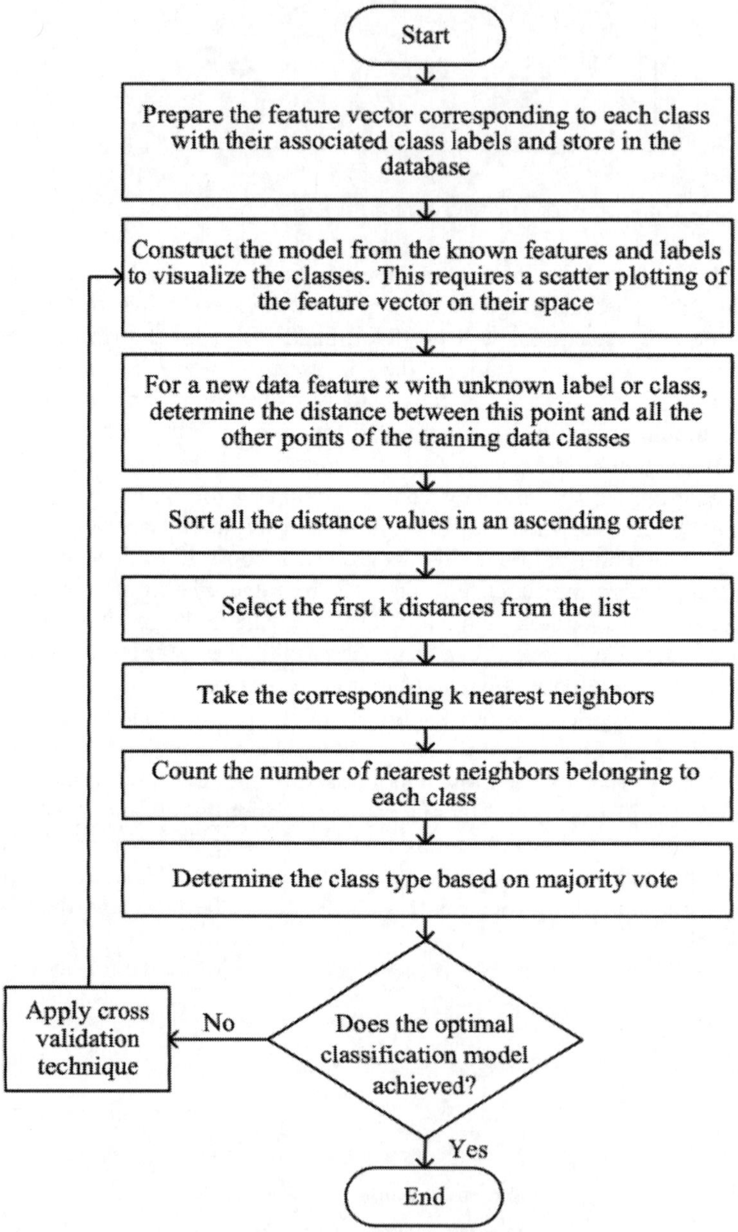

Figure. 4. Flowchart of the majority vote based KNN classification procedure.

3 Case studies

Seven component faults in a two-shaft industrial gas turbine engine were considered as shown in Table 1. In order to generate the required data for training and test, at first, the performance prediction model was developed using a simulation program, tuned to represent a gas turbine which is thermodynamically similar to GE LM2500 engine. It consists of a 16-stage axial compressor (C), a 2-stage compressor turbine (CT) and a 6-stage power turbine (PT) that can produce 23.3 MW power output. Figure 5 illustrates the schematics of the basic components along with the gas-path measurements used for the detection.

Table 1. Description of the considered component faults

Fault Type	Description
Fault 1	Compressor fouling (CF)
Fault 2	CT erosion (CTE)
Fault 3	PT erosion (PTE)
Fault 4	CF and CTE
Fault 5	CF and PTE
Fault 6	CTE and PTE
Fault 7	CF, CTE, and PTE

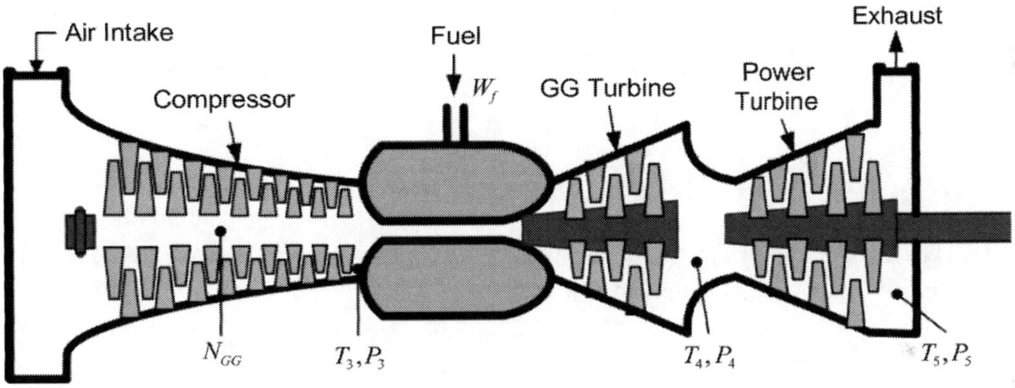

Figure. 5. Two-shaft gas turbine schematics.

The necessary clean condition data is generated taking into account ambient condition variations ranging from -45 °C to 50 °C for ambient temperature and from 0 to 3% for ambient pressure. In order to generate the test data that can represent a deteriorated engine operating condition, artificial fault case patterns derived from the assumed seven component fault scenarios were implanted into the model and simulated. Overall 20955 data samples (7620 no-fault and 13335 fault data samples (1905 samples drawn from each fault type) were used to train and test the system. For compressor fouling, flow capacity losses ranging from 0 to -5% and isentropic efficiency losses ranging from 0 to -2.5% were considered in different ratios. Similarly, for turbine erosion, flow capacity changes ranging from 0 to 5% and isentropic efficiency changes ranging from 0 to -2.5% were considered. In the end, in order to satisfy the real measurement situations, a zero mean Gaussian white noise is superimposed to the data. Table 2 presents the maximum noise-level considered in standard deviation values of each gas path measurement.

Table 2. The sensors' noise standard deviations in %

Sensor	T3	P3	NGG	Wf	T4	P4	T5	P5
Noise (±σ)	0.4	0.25	0.05	0.5	0.4	0.25	0.4	0.25

4 Results and discussion

4.1 Noise reduction

In order to select the optimal AANN structure, hidden layers with different number of neurons were considered (for the mapping and de-mapping layers from 8 to 50 neurons and for the bottleneck layer from 1 to 7 neurons). A large number of neurons were required in the mapping and de-mapping layers because these layers are quite complex non-linear functions. In the training process the learning data sample was divided into three groups: training, validation, and test datasets in different rations. Taking into account the model complexity and training computational efficiency, the model structure 8:30:4:30:8 was selected as the best structure among the other potential structures. The selected structure was able to provide a noise reduction of over 85% for most of the measurements.

Owing to space limitations only the denoising results for the temperature measurements are presented as shown in Figures 6-8. The difference between noisy input data and network output data refers to the level of noise reduction by the network. In general, one can see that a significant level of noise has been removed from the data corresponding to each measurement although the effectiveness of the network is not uniform between different measurements. Eq. (6) shows the formula that has been used to compute the percentage noise reduction for each parameter.

$$\text{Noise reduced (\%)} = \frac{1}{N}\sum_{i=1}^{N}\frac{\left[(X_n)_i - (X_o)_i\right]}{\left[(X_n)_i - (X_t)_i\right]}\times 100 \qquad (6)$$

Where Xn is the noisy data input, Xo is the de-noised AANN output, X_t is the target, i is the ith sample data, and N is the number of samples.

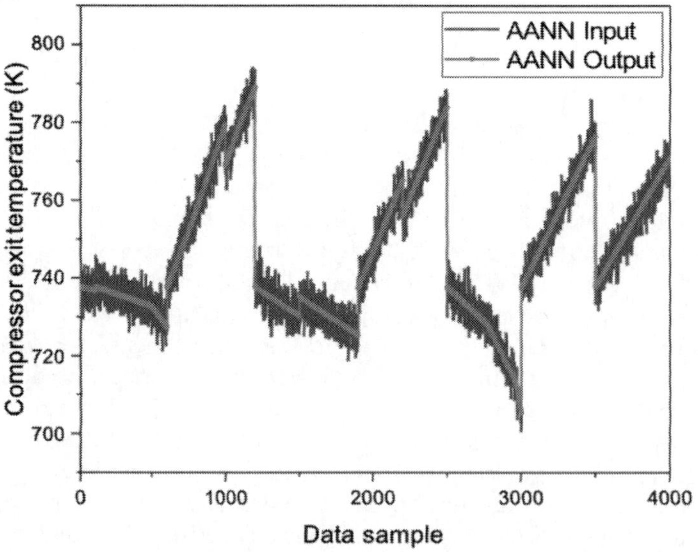

Figure. 6. Noise removal results of 4000 data examples for compressor exit temperature.

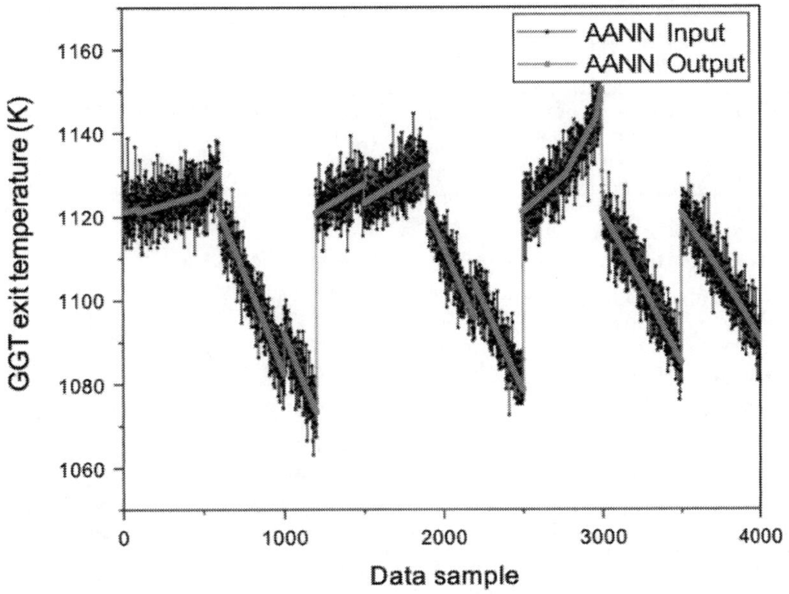

Figure. 7. Noise removal results of 4000 data examples for gas generator turbine exit temperature.

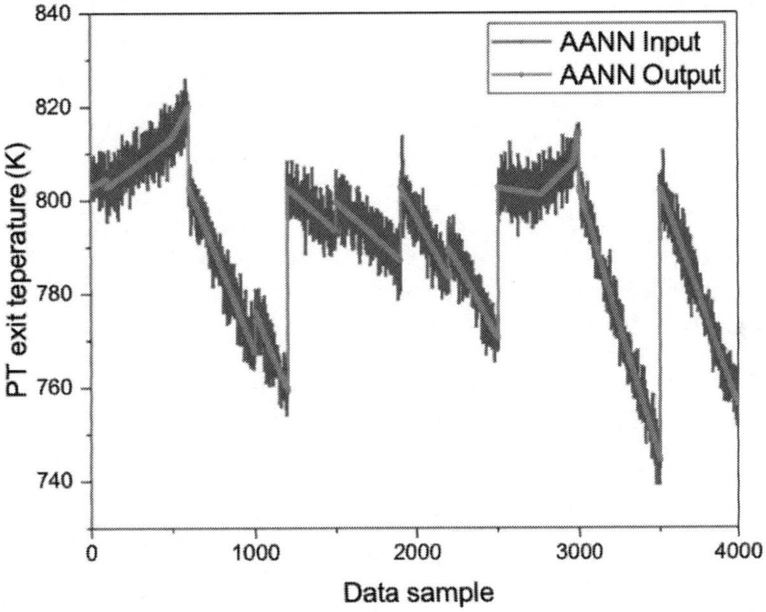

Figure. 8. Noise removal results of 4000 data examples for power turbine exit temperature.

In order to see the influence of noise level on the effectiveness of the de-noising network, noise values of $\pm 1\sigma$, $\pm 2\sigma$, and $\pm 3\sigma$ were considered. Table 3 shows the percentage noise reductions obtained. In fact, as the level of the noise increases, the magnitude of the noise to be reduced from each measurement decreases. Nevertheless, for all the measurements a significant amount of noise has been removed. Since the AANN output is the input of the KNN classifier, the remaining percentage of noise in the input data helps to enhance the flexibility/robustness of the classifier when unexpected data features come into the system.

Table 3. Effectiveness of the AANN at different noise-levels.

Noise Level	Percentage noise reduction								
	T3	P3	N_{GG}	Wf	T4	P4	T5	P5	Average
±1σ	87.6	85.8	86.7	84.5	87.5	84.6	86.7	86.3	86.21
±2σ	81.5	80.7	82.8	81.3	81.4	80.6	79.3	81.5	81.14
±3σ	77.3	76.4	73.7	74.9	77.5	73.5	74.5	77.1	75.61

4.2 Fault detection

In this section, the fault detection performance of the KNN technique was evaluated using the generated performance data that can represent both the healthy and faulty engine operating conditions. In order to select the appropriate classifier model, effects of the major controlling factors such as the distance function, the number of nearest neighbors, and training data size were investigated. A total of 20955 patterns (7620 no-fault and 13335 fault patterns) were used to train and test the model. The healthy operating conditions were assigned as positive (+), while the faulty conditions were assigned as negative (-). The detection accuracy (i.e., how accurately the two data classes are distinguished) was computed based on the standard detection decision matrix approach.

To analyse the effectiveness of different KNN functions and select the most suitable function for our data, medium KNN, coarse KNN, cosine KNN, cubic KNN, and weighted KNN functions were considered. These classifiers are different, to some extent, in their classification algorithm, computational speed, memory requirement, flexibility, and interpretability. Table 4 illustrates the classification results obtained. The maximum accuracy is provided by the Fine KNN model (98.3%), while the minimum by Coarse KNN (93.9%). The classification accuracy using Weighted KNN and Cosine KNN is 0.6% and Medium KNN and Cubic KNN is 1% lower than the maximum accuracy. As per this table, the Fine KNN was chosen for further investigation. Since a very low-level faults were also considered, as can be seen in the table, some patterns were recognized by the classifier as a fault-free patterns. An attempt to minimize the number of missed detection may increase false detections, and vice versa. However, since too early maintenance is not economical, it is better to have missed detections than false alarms.

Table 4. Classification performance of different KNN functions.

KNN Type	TPs	FPs	TNs	FNs	MD	FA	ODA
Fine KNN	12959	0	7620	375	2.8	0	98.3
Medium KNN	12858	0	7620	477	3.6	0	97.3
Coarse KNN	12294	0	7620	1044	7.8	0	93.9
Cosine KNN	12948	0	7620	387	2.9	0	97.7
Cubic KNN	12872	0	7620	463	3.6	0	97.3
Weighted KNN	12947	0	7620	388	2.9	0	97.7

4.2.1 Effect of distance matric and weigh

In order to investigate whether the classification effectiveness is influenced by the distance function and the distance weight or not, five distance matrices with three different weight functions were considered. For this purpose, the Fine KNN model, the KNN with better classification performance as reported in Table 4, was selected. Table 5 presents the overall detection accuracy of this model for the possible combinations of the distance matric and weight functions. As seen in this result, the influence of the distance matric and the weight function was insignificant. Therefore, the Euclidian function was chosen as the distance matric for our model, and it is also one of the most commonly used functions in the literature.

Table 5. Effect of the distance matric and distance weight on the classification performance.

Distance Matric	Distance Weight	ODA
Euclidian	Equal	98.3
	Inverse	98.3
	Square inverse	98.3
Minkowski	Equal	97.7
	Inverse	98.0
	Square inverse	98.1
Manhattan	Equal	98.2
	Inverse	98.3
	Square inverse	98.3
Cosine	Equal	98.2
	Inverse	98.2
	Square inverse	98.2
Correlation	Equal	97.8
	Inverse	98.0
	Square inverse	98.1

4.2.2 Effect of training data size

The accuracy of machine learning methods is influenced by the number of samples used for training. Small training data samples may cause overfitting and high prediction errors. Conversely, regardless of memory and computational requirements, for practical concerns, a more training data size often provides a better testing accuracy. However, after a certain limit, the effect of increasing the amount of training data samples will be negligible [28]. A model that is overfitting will be less good for a new dataset than a model with a good fit. One way to have a look into generalizability is to have a training set and test set. In this paper, one of the most widely used approaches known as the k fold cross-validation technique was applied. According to this approach, the learning dataset is divided into k equal groups. During the learning process, the training is done on the k-1 data groups and test on the rest group, and so on for all possible combinations. Accordingly, to examine the sensitivity of the KNN classifier model towards the number of folds, six different cases (5, 10, 15, 20, 25 and 30 folds) were taken into account. Table 6 presents the fault detection results obtained. In general, since the number of cross-validation samples was high enough, in all the six cases the classifier provided over 98% accuracy. Consequently, the 10 fold cross-validation technique, which is also the most commonly used validation approach in different KNN applications [29], has been selected.

Table 6. Effect of training data size on the classification accuracy.

Number of folds	Training time (s)	MD	FA	ODA
5	2	2.9	0	98.3
10	2	2.9	0	98.3
15	2	2.9	0	98.3
20	3	2.9	0	98.3
25	2	2.9	0	98.3
30	4	2.9	0	98.3

To summarize, as we can see from the sensitivity analysis, the effect of most of the controlling factors on the classification performance of the KNN model was insignificant. The fault detection accuracy mainly depends on the relationship between the low-level faults and the measurement noise values. As a mature of fact, when the fault signatures associated with these conditions are similar, there will be a possibility of miss detection and/or false alarms. However, in the present work, due to the de-noising process prior to the fault detection, a reasonably high-degree of accuracy was obtained.

5 Conclusions

In this paper, a combined AANN and KNN technique is devised for a gas turbine engine noise reduction and fault detection application. The AANN structure is used to minimize the measurement noise prior to the fault detection process. Whereas the KNN technique is used to determine whether the engine is healthy or unhealthy. The data required to demonstrate and evaluate the performance of the developed method is generated from a model simulation, tuned to represent a two-shaft gas turbine which is thermodynamically similar to LM2500 engine. As gas turbine physical problems, compressor fouling and turbine erosion, which are the major and usual causes of gas turbine gas path component(s) performance deterioration, has been considered. White Gaussian noise is added to the performance data in order to account the measurement noise. The test results showed that the AANN algorithm was able to reduce over 85% of the measurement noise while the KNN classifier was able to detect 98.3% of the faulty conditions successfully. The proposed method can, therefore, be employed for gas turbine gas path fault detection.

Nomenclature

AANN	Autoassociative neural network
ANN	Artificial neural network
C	Compressor
CF	Compressor fouling
CT	Compressor turbine
CTE	Compressor turbine erosion
EKF	Extended Kalman filter
FD	False detection
FDI	Fault detection and isolation
FDII	Fault detection isolation and identification
FL	Fuzzy logic
FN	False negative
FP	False positive
GPA	Gas path analysis
KF	Kalman filter
MD	Missed detection
MFs	Membership function
MLP	Multilayer perceptron
ODA	Overall detection accuracy
PNN	Probabilistic neural network
PT	Power turbine
PTE	Power turbine erosion

RFP	Rate of false positive
RMS	Root mean square
RTP	Rate of true positive
SVM	Support vector machine
TN	True negative
TP	True positive

Acknowledgments

This research was partially funded by the Knowledge Foundation (KKS) under the project DIAGNOSIS.

References

1. A. D. Fentaye, A. T. Baheta, S. I. Gilani, and K. G. Kyprianidis, "A Review on Gas Turbine Gas-Path Diagnostics: State-of-the-Art Methods, Challenges and Opportunities," Aerospace, vol. 6, p. **83**, (2019).
2. Y. Qingcai, S. Li, Y. Cao, and N. Zhao, "*Full and Part-Load Performance Deterioration Analysis of Industrial Three-Shaft Gas Turbine Based on Genetic Algorithm*," In Proceedings of the ASME Turbo Expo 2016: Turbomachinery Technical Conference and Exposition, Harbin, China, (13–17 June 2016); p. V006T005A016.
3. T. Kobayashi and D. L. Simon, "Evaluation of an enhanced bank of Kalman filters for in-flight aircraft engine sensor fault diagnostics," J ENG GAS TURB POWER, vol. 127, pp. 497-504, (2005).
4. A. J. Volponi, H. DePold, R. Ganguli, and C. Daguang, "The Use of Kalman Filter and Neural Network Methodologies in Gas Turbine Performance Diagnostics: A Comparative Study," J ENG GAS TURB POWER, vol. 125, pp. 917-924, (2003).
5. B. Pourbabaee, N. Meskin, and K. Khorasani, "Sensor Fault Detection, Isolation, and Identification Using Multiple-Model-Based Hybrid Kalman Filter for Gas Turbine Engines," IEEE Transactions on Control Systems Technology, (2015).
6. R. Ganguli, *Gas Turbine Diagnostics: Signal Processing and Fault Isolation*: CRC press: Boca Raton, FL, USA, (2012).
7. F. Lu, Y. Chen, J. Huang, D. Zhang, and N. Liu, "*An integrated nonlinear model-based approach to gas turbine engine sensor fault diagnostics*," Proc IMechE Part G: J Aerospace Engineering, vol. **228**, pp. 2007-2021, (2014).
8. F. Lu, T. Gao, J. Huang, and X. Qiu, "A novel distributed extended Kalman filter for aircraft engine gas-path health estimation with sensor fusion uncertainty," Aerospace Science and Technology, vol. **84**, pp. 90-106, (2019).
9. Q. Yang, S. Li, and Y. Cao, "Multiple model-based detection and estimation scheme for gas turbine sensor and gas path fault simultaneous diagnosis," JMST, vol. 33, pp. 1959-1972, (2019).
10. V. Zaccaria, M. Stenfelt, I. Aslanidou, and K. G. Kyprianidis, "Fleet Monitoring and Diagnostics Framework Based on Digital Twin of Aero-Engines," in ASME Turbo Expo 2018: Turbomachinery Technical Conference and Exposition, Oslo, Norway (June 11-15, 2018).
11. M. Rahman, V. Zaccaria, X. Zhao, and K. Kyprianidis, "Diagnostics-Oriented Modelling of Micro Gas Turbines for Fleet Monitoring and Maintenance Optimization," Processes, vol. **6**, p. 216, (2018).

12. D. Zeng, D. Zhou, C. Tan, and B. Jiang, "Research on Model-Based Fault Diagnosis for a Gas Turbine Based on Transient Performance," Applied Sciences, vol. **8**, p. 148, (2018).
13. M. A. Kramer, "Nonlinear Principal Component Analysis Using Autoassociative Neural Networks," AIChE Journal, vol. **37**, p. 233, (1991).
14. M. A. Kramer, "Neutral network applications in chemical engineeringAutoassociative neural networks," COMPUT CHEM ENG, vol. **16**, pp. 313-328, 1992/04/01 (1992).
15. P. J. Lu and T. C. Hsu, "Application of autoassociative neural network on gas-path sensor data validation," J PROPUL POWER, vol. **18**, pp. 879-888, (2002).
16. R. Ganguli and B. Dan, "Trend Shift Detection in Jet Engine Gas Path Measurements Using Cascaded Recursive Median Filter With Gradient and Laplacian Edge Detector," J ENG GAS TURB POWER, vol. **126**, pp. 55-61, (2004).
17. S. Sina Tayarani-Bathaie and K. Khorasani, "Fault detection and isolation of gas turbine engines using a bank of neural networks," J PROCESS CONTR, vol. **36**, pp. 22-41, 12// (2015).
18. M. Amozegar and K. Khorasani, "An ensemble of dynamic neural network identifiers for fault detection and isolation of gas turbine engines," Neural Networks, vol. **76**, pp. 106-121, 4// (2016).
19. Z. N. Sadough Vanini, N. Meskin, and K. Khorasani, "Multiple-Model Sensor and Components Fault Diagnosis in Gas Turbine Engines Using Autoassociative Neural Networks," J ENG GAS TURB POWER, vol. **136**, p. 091603, (2014).
20. I. Loboda and M. A. Olivares Robles, "Gas turbine fault diagnosis using probabilistic neural networks," INT J TURBO JET ENG., vol. **32**, pp. 175-191, (2015).
21. A. D. Fentaye, A. T. Baheta, and S. I. U.-H. Gilani, "Gas turbine gas-path fault identification using nested artificial neural networks," Aircraft Engineering and Aerospace Technology, vol. **90**, pp. 992-999, (2018).
22. D. Amare, T. Aklilu, and S. Gilani, "Gas path fault diagnostics using a hybrid intelligent method for industrial gas turbine engines," J. Braz. Soc. Mech. Sci. & Eng, vol. **40**, p. 578, (2018).
23. W. Donat, K. Choi, W. An, S. Singh, and K. Pattipati, "Data visualization, data reduction and classifier fusion for intelligent fault diagnosis in gas turbine engines," J ENG GAS TURB POWER, vol. **130**, p. 041602, (2008).
24. I. Loboda, "*Gas turbine fault classification using probability density estimation*," in Proceedings of ASME Turbo Expo: Turbine Technical Conference and Exposition, GT2014-27265, Germany, Dusseldorf, (2014).
25. A. Fentaye, V. Zaccaria, M. Rahman, M. Stenfelt, and K. Kyprianidis, "*Hybrid model-based and data-driven diagnostic algorithm for gas turbine engines*," Proceedings of ASME Turbo Expo 2020: Turbomachinery Technical Conference and Exposition, GT2020-14481, (2020).
26. A. D. Fentaye, S. I. Ul-Haq Gilani, A. T. Baheta, and Y.-G. Li, "Performance-based fault diagnosis of a gas turbine engine using an integrated support vector machine and artificial neural network method," Proceedings of the Institution of Mechanical Engineers, Part A: Journal of Power and Energy, vol. **233**, pp. 786-802, (2019).
27. P. J. Lu, M. C. Zhang, T. C. Hsu, and J. Zhang, "An evaluation of engine faults diagnostics using artificial neural networks," J ENG GAS TURB POWER, vol. **123**, pp. 340-346, (2001).

28. R. Bettocchi, M. Pinelli, P. R. Spina, and M. Venturini, "Artificial Intelligence for the Diagnostics of Gas Turbines—Part I: Neural Network Approach," J ENG GAS TURB POWER, vol. **129**, pp. 711-719, (2006).
29. S. Borra and A. D. Ciaccio, "Measuring the prediction error. A comparison of cross validation, bootstrap and covariance penalty methods," Computational Statistics & Data Analysis, vol. **54**, pp. 2976-2989, (2010).

Perspectives on Health and Usage Monitoring Systems (HUMS) of helicopters

Alexandre Mauricio[1,2], *Junyu* Qi[1,2], *Linghao* Zhou[3], *Wenyi* Wang[4], *David* Mba[5], and *Konstantinos* Gryllias [1,2,*]
[1]Department of Mechanical Engineering, KU Leuven, 3001, Leuven, Belgium
[2]Dynamics of Mechanical and Mechatronic Systems, Flanders Make, 3001, Leuven, Belgium
[3] London South Bank University, School of Engineering, SE1 0AA, London, UK
[4] Defence Science and Technology Organization (DSTO), Air Vehicles Division, Melbourne, VIC 3011, Australia
[5] De Montfort University, Faculty of Technology, LE1 9BH, Leicester, UK

Abstract. Helicopters are extensively used in civil applications as they are versatile in their capabilities to manoeuvre. Their operation under harsh conditions and environments demand for a strict maintenance plan. Main gearboxes (MGB) of helicopters are a critical component responsible for reducing the high input speed generated from the gas turbine engines. Health and Usage Monitoring Systems (HUMS) are installed in an effort to monitor the health state of the transmission systems, and ideally, to detect and diagnose the type of a generated fault. Even though the development of HUMS contributed to the reduction of worldwide helicopter accident rate, more advanced systems are needed based on the investigation of the air accidents of AS332 L2 Super Puma in Scotland in 2009 and of EC225 LP Super Puma in Bergen in 2016, due to failure of a planet gear of the MGB. A plethora of signal processing methodologies have been proposed for the early detection of faults but often they fail in complex structures, such as planetary gearboxes operating under various conditions. In this paper the performance of a recently proposed diagnostic tool, called IESFOgram, is evaluated and compared with state of the art techniques, applied on signal captured on a Category A Super Puma SA330 MGB.

1 Introduction

Health and Usage Monitoring Systems (HUMS) conduct helicopter health monitoring by extracting Health Indicators (HI) from acquired vibration data. However, HI's have a wide range of values depending on a plethora of conditions of the helicopter besides their health state (e.g. temperature, speed, torque, etc.), making the implementation of specific thresholds of the HI often unreliable for the health monitoring. Analysis of the time-domain of the vibration data is also non-practical, as MGB of helicopters are complex machinery with many sources. Analysis of the frequency-domain provides a more direct way to monitor the health condition of its components by tracking the amplitude at specific frequencies related to specific components. Signal processing techniques are usually applied in this scenario in

* Corresponding author: konstantinos.gyllias@kuleuven.be

order to detect modulations related to the fault that may be hidden in the spectrum [1,2]. One of the most common approaches in bearing fault detection is the Envelope Analysis, where the signal is demodulated in order to detect the lower cyclic modulations of the bearing. This method is sensitive to the noise, so it is commonly used alongside with a band pass filter around the band of frequencies excited by the impulses of the bearing fault. The selection of the band pass filter can be made mainly by two general approaches: exploitation of engineering knowledge and experience; or through band selection tools. The band selection tools provide the advantage of (semi-) automated selection without requiring expertise and/or trial-and-error of band-pass filtering. The most widely used method is the Fast Kurtogram, which maximises the value of Spectral Kurtosis of a series of filters. However, there are other known phenomena that generate high Kurtosis values, resulting in the selection of incorrect bands [3]. In the frames of the analysis of vibration data of a MGB with damaged bearing, Zhou et. al. [1] concluded that the Kurtogram was ineffective by itself, and the damage was only diagnosable after application of a self-adaptive noise cancellation filter.

Machine learning based method have also seen an increasing interest for damage detection of rotating machinery due to their capabilities of classifying signals under blind assumptions [4,5]. Usually a set of features, statistical features for example, are used as inputs to train and test the model. The main disadvantage is that they need a high number of signals (samples) in order to properly train the classification model.

The last decade, the interest in the development and use of cyclostationary based tools for condition monitoring, such as the Cyclic Spectral Correlation (CSC) and the Cyclic Spectral Coherence (CSCoh), has been increased as they achieve high performance in revealing hidden periodicities in their bi-variable representations. Direct analysis of these results is difficult to perform, unless the analyst has extensive prior experience. On the other hand, the integration of the spectral frequency variable results in a one dimensional envelope spectrum, the Enhanced Envelope Spectrum (EES), detailing the cyclic modulations present in the signal. Furthermore, when the integration is performed on a specific narrow band instead of the full band, only the signature with carriers inside the integrated band are kept in the spectrum, nominated as the Improved Envelope Spectrum (IES). This method achieves robust results in extracting hidden modulations in the spectrum, and results in a classical Envelope Spectrum of easy analysis [8,9]. Recently, the IESFOgram (Improved Envelope Spectrum via Feature Optimization –gram) has been proposed as an automated band selection tool for application on the CSC and CSCoh bi-variable maps, leading to robust results in the field of bearing diagnostics [9].

The key objective of this paper is to evaluate the performance of the IESFOgram and compare it with state of the art methods by applying it on real vibration data acquired from a MGB of a Super Puma under various levels of damage on the planet bearing, operating under different speed and load conditions.

2 Band selection tools theory

2.1 Fast Kurtogram

Fast Kurtogram is a fourth-order spectral analysis tool which has been introduced for detecting and characterizing non-stationarities in a signal and it has been widely used as an automated band selection tool based on the maximization of Spectral Kurtosis (SK). Therefore a 1/3 binary tree filter bank structure is prepared and the SK of each band is estimated. The band where the SK is maximized corresponds to an optimal filter band which presents a high Signal-to-Noise Ratio (SNR) leading to a Squared Envelope Spectrum (SES) after demodulation, where the fault harmonics are enhanced. On the other hand it has been proven that Kurtogram is sensitive to outliers.

2.2 IESFOgram

Rolling element bearing signals can be described as second-order cyclostationary signals, due to the slippage of elements, presenting a periodic autocorrelation of period T:

$$R_{2x}(\tau, t) = \mathbb{E}\{\, x(t)x(t-\tau)^{*}\} = R_{2x}(\tau, t+T) \quad (1)$$

where x(t) is the time signal, $\mathbb{E}$ is the ensemble average and τ is the time-lag. Cyclostationary based tools, such as Cyclic Spectral Correlation (CSC) and Coherence (CSCoh) have received an ever increasing interest due to their ability in extracting hidden modulations. CSC estimates the correlation between a modulating frequency (the cyclic frequency α) and a carrier frequency (the spectral frequency f) presented in a bi-variable map [7]:

$$CSC(\alpha, f) = \lim_{T\to\infty} T^{-1}\mathbb{E}[\mathcal{F}(x(t))^{*}\mathcal{F}(x(t-\tau))] \quad (2)$$

where F(x(t)) is the Fourier transform of the signal x(t), and $\mathbb{E}$ is the ensemble average.

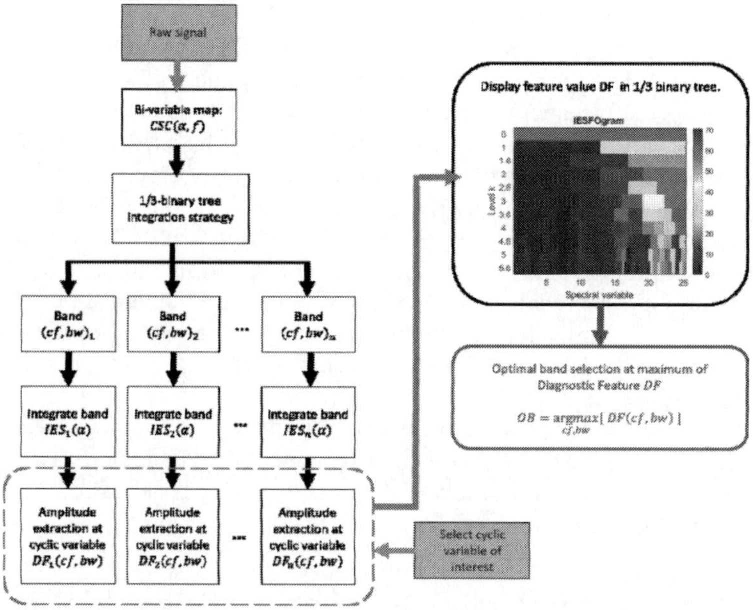

Figure. 1. Schematic description of the IESFOgram procedure.

The Cyclic Spectral Coherence (CSCoh) is the whitened version of the CSC:

$$CSCoh(\alpha, f) = \frac{CSC(\alpha,f)}{\sqrt{CSC(0,f).CSC(0,f-\alpha)}} \quad (3)$$

The integration of the bi-variable maps over a band [F_1, F_2] of the spectral frequency *f* leads to the Improved Envelope Spectrum (IES) and to the Enhanced Envelope

Spectrum (EES) ($F_1 = 0$, $F_2 = F_s/2$, F_s is the sampling frequency) which are equivalent to a demodulation spectrum:

$$IES(\alpha) = \frac{1}{F_2 - F_1} \int_{F_1}^{F_2} | CSCoh(\alpha, f) | df \quad (4)$$

Recently the IESFOgram has been proposed as a tool is order to automatically select the optimal integration band, similarly to the Kurtogram [8]. A series of integrating iterations for different combinations of bandwidth and centre frequency are performed following a 1/3 binary tree. A diagnostic feature (DF), equal to the sum of the amplitude of k=3 harmonics of the bearing fault frequency ffault normalized by the surrounding noise level (estimated at a band [f_{fault}-/+1/4f_{shaft}]), is estimated at each band, as described by Eq. (5).

$$DF = \sum_{k=1}^{3} \frac{IES(k \times \alpha_{fault})}{\frac{1}{2f_b}[\int_{k\alpha_{fault}-f_b}^{k\alpha_{fault}-f_b} IES(\alpha)d\alpha - IES(k \times \alpha_{fault})]} \quad (5)$$

The highest feature value corresponds to the optimal integration band as described in Figure 1.

3 Experimental Setup

A Category A Super Puma SA330 main Gearbox (MGB) has been used in order to investigate the accuracy of planetary bearing defect diagnosis methods. The measurements have been acquires from a dedicated test rig, presented in Fig. 2, in the frames of a project entitled "MGH - Helicopter main gearbox health" sponsored by the European Aviation Safety Agency (EASA).

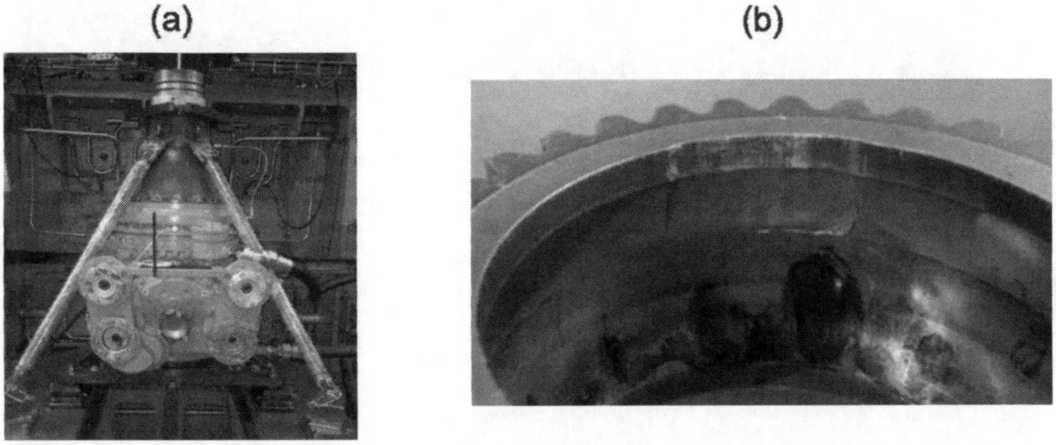

Figure. 2. (a) Tested Main Gearbox, (b) Major fault seeded on the planet outer race.

The MGB has several stages of reduction, parallel and planetary. Bearing defects have been seeded in the outer race of the planetary bearing at the second epicyclic module. More precisely, the outer race faults have been seeded on the inner bore of the planet gear, which is the outer race of the bearing. The planet gears are driven by the 2nd epicyclic sun gear and the ring gear.

4 Results

For each case, different number of accelerometers were used, but the authors show the results in this section for the accelerometer signal that is closest to the damage source and that is not corrupted or truncated.

4.1 Case under 23,000 rpm and 1760 kW operating conditions

The vibrations signals analysed in this first case were captured under 23,000 rpm shaft speed and 1760 kW. The signals were also measured for three cases health conditions of the outer race of the planet bearing of the second epyciclic stage: healthy, minor fault, and major fault.

The minor fault is approximately 10mm wide, and the major fault is approximately 30mm. Based on the kinematics of the main gearbox, the frequency of interest can be calculated to be the BPFO, with a cyclic frequency of 97.4 Hz. Detection of this component and its harmonics results in the diagnosis of the damage.

The SES and EES were also extracted and analysed, however no peaks of the BPFO were detected in any of the cases.

The Fast Kurtogram for each health condition, seen in Fig. 3, were used to select the band for filtering that results in the SES based on Fast Kurtogram seen in Fig. 4. Low kurtosis values are represented in blue, and the optimal band of each Fast Kurtogram is represented in yellow as the highest kurtosis value.

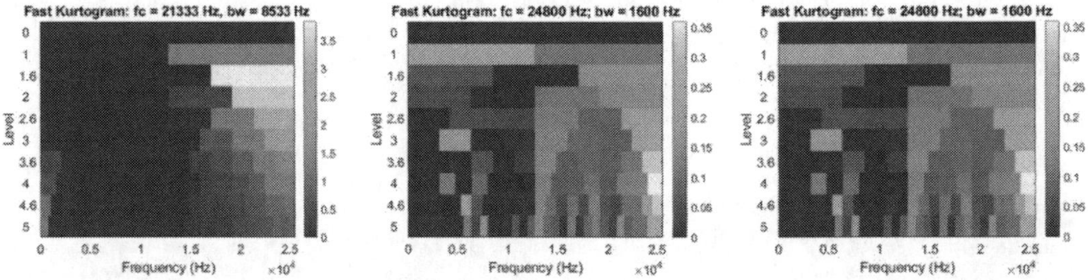

Figure. 3. Fast Kurtogram of: (left) healthy case, (middle) minor fault case, and (right) major fault case

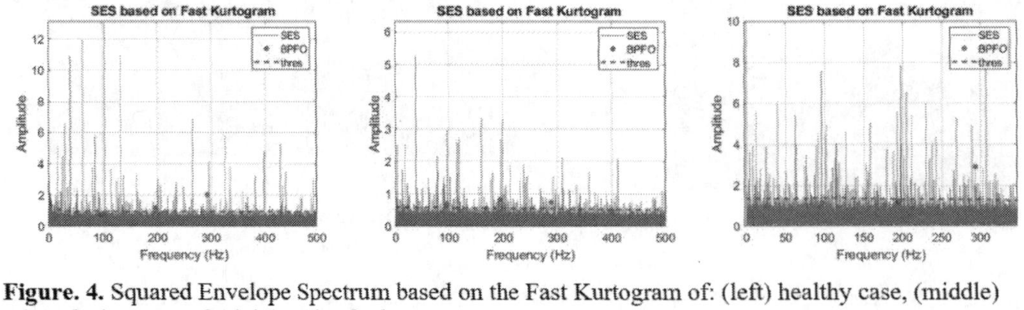

Figure. 4. Squared Envelope Spectrum based on the Fast Kurtogram of: (left) healthy case, (middle) minor fault case, and (right) major fault case.

The harmonics of the BPFO cannot be detected with confidence above the noise level, as can be seen on the spectra.

On the other hand, the IESFOgram for the same signals, seen in Fig. 5, and its corresponding IES, seen in Fig. 6, detect the peaks at the BPFO's harmonics well above the noise level for the damaged cases, and thus providing a robust diagnosis.

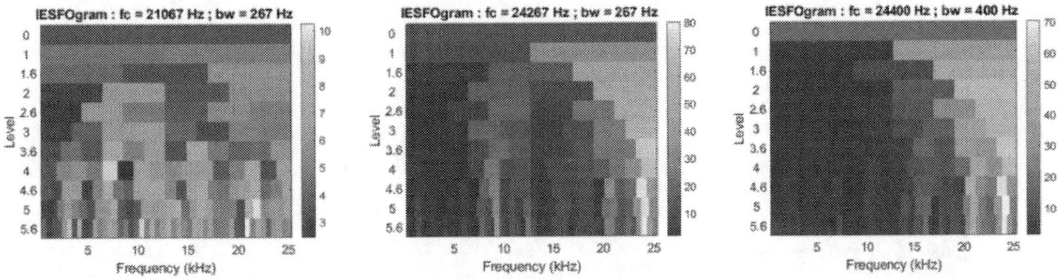

Figure. 5. IESFOgram of: (left) healthy case, (middle) minor fault case, and (right) major fault case

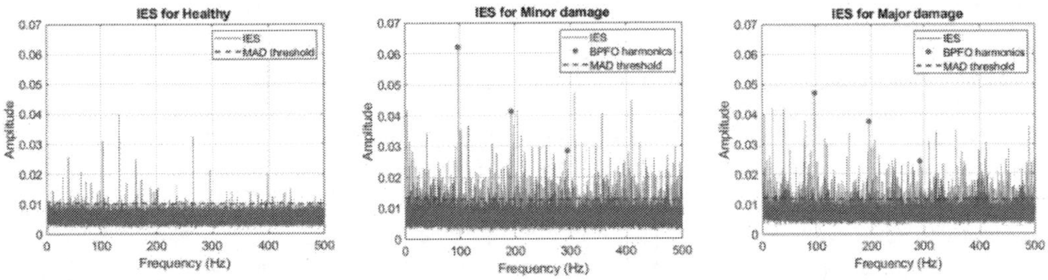

Figure. 6. Improved Envelope Spectrum based on the IESFOgram of: (left) healthy case, (middle) minor fault case, and (right) major fault case.

4.2 Case under 16,000 rpm and 180 kW operating conditions

On the second test case, the operating conditions were lower than on the first case, at 16,000 rpm rotation speed and 180 kW. The minor fault has a length, width and depth of 5mm, 3mm, and 1.5mm respectivelly. The major fault has in turn a length, width and depth of 8mm, 6mm, and 2.5mm. Under these speed conditions, the BPFO is 68.5 Hz.

The peaks at the BPFO harmonics were not detectable for any health condition for the SES, the EES, and the SES based on the Fast Kurtogram, as seen in Fig. 7.

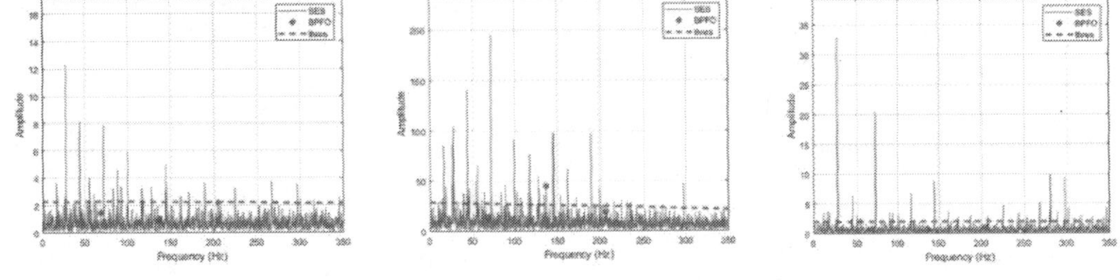

Figure. 7. Squared Envelope Spectrum based on the Fast Kurtogram of: (left) healthy case, (middle) minor fault case, and (right) major fault case.

The IES based on the IESFOgram exhibits the first harmonic of the BPFO well above the noise level, however the peak at the BPFO is also detected in the healthy case. The amplitude increases with the severity level, which can provide a diagnosis of increasing damage, but the detection of the BPFO at the healthy case does not provide full confidence in the diagnosis. The authors can infer that the peak at the 68.5 Hz in the healthy case can be due to the damage of the outer race during the bearing mounting in the setup, the ball pass frequency due to varying stiffness from the balls in contact, or other component altogether [10].

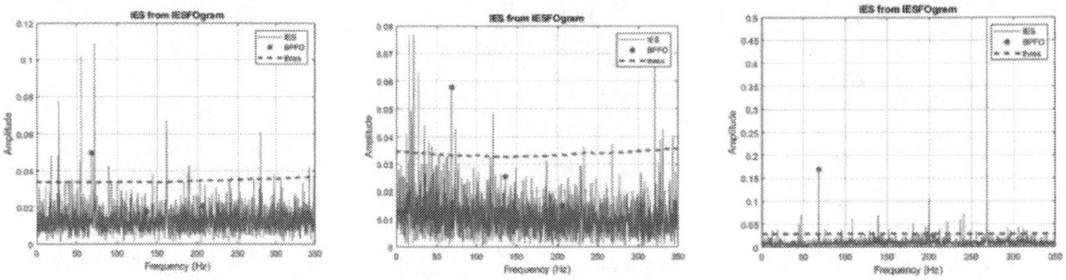

Figure. 8. Improved Envelope Spectrum based on the IESFOgram of: (left) healthy case, (middle) minor fault case, and (right) major fault case.

4.3 Case under 23,000 rpm and 1300 kW operating conditions

The last case was only captured for two health conditions: healthy and major damage on the outer race of the bearing. The setup was under 23,000 rpm and 1300 kW operating conditions on both situations.

On this case, the peaks at the 97.4 Hz are more easily extracted when compared to the previous cases, as the BPFO can be detected not only on the IES based on the IESFOgram, ass seen in Fig. 9, but also with the classical SES and EES. The only exception is the SES based on the Fast Kurtogram, where the BPFO harmonics are not detectable on the major fault spectrum.

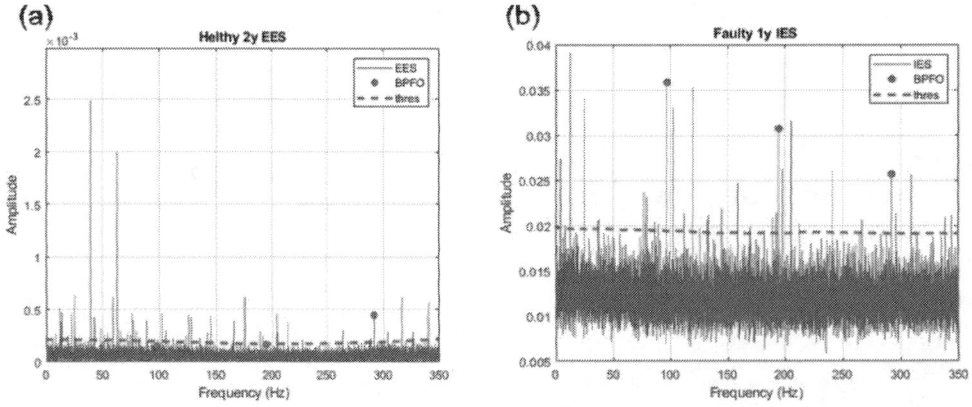

Figure. 9. Improved Envelope Spectrum based on the IESFOgram of: (left) healthy case, and (right) major fault case.

4.4 Comparison of methodologies

The results of the methodologies' comparison are briefly presented in Table 1 for the three analysed cases. The "NO" marks the NO fault detection while the "YES" highlights the fault detection. The green colour corresponds to the correct result while the red colour corresponds to an error. The Fast Kurtogram based SES shows low performance, either because the signal of interest is well masked by other components, or because an incorrect filtering band is selected. Moreover the diagnosis by the classical SES and EES analysis is successful for only one of the three cases. The integration of the CSCoh using the band selected from the IESFOgram results in spectra where the BPFO is clearly identified in all the damaged cases, however it is also extracted on the healthy signal of one of the cases. Based on Table 1 it can be concluded that the IES estimated based on the IESFOgram achieves the best results compared to the other 3 state of the art methodologies. On the other hand, one false alarm is present and the authors are now working on this issue in order to clarify the reason of appearance and the improvement of the general methodology targeting towards a robust diagnostic method with a limited rate of false alarms and missed detections.

Table. 1. Detection of the BPFO for all cases for different signal processing methods.

Operating conditions	Damage	SES	EES	IES	Fast Kurtogram
Speed 23,000 rpm Load 1760 kW	Healthy	NO	NO	NO	NO
	Minor Damage	NO	NO	YES	NO
	Major Damage	NO	NO	YES	NO
Speed 23,000 rpm Load 1760 kW	Healthy	NO	NO	YES	NO
	Minor Damage	NO	NO	YES	NO
	Major Damage	NO	NO	YES	NO
Speed 23,000 rpm Load 1760 kW	Healthy	NO	NO	NO	NO
	Major Damage	YES	YES	YES	NO

5 Conclusions

In this paper, the IES based on the IESFOgram is proposed for diagnosis of bearing diagnostics of helicopter's main gearbox. The procedure is applied over signals captured on a Category A Super Puma SA330 helicopter main gearbox transmission, for various speed conditions, load conditions and severity levels of damage on a rolling element bearing. The procedure is compared against the classical SES, the EES and the SES based on the Fast Kurtogram. The detection of the BPFO for all cases and the mentioned methods are summarized in Table 1.

The Fast Kurtogram based SES shows low performance, either because the signal of interest is well masked by other components, or by selecting the incorrect band for filtering. Diagnosis by the classical SES and EES analysis is robust for one of the three cases. The integration of the CSCoh using the band selected from the IESFOgram results in spectra where BPFO are clearly identified in all the damaged cases, however the BPFO it is also extracted on the healthy signal of one of the cases. Therefore early detection of the faults can be extracted with the use of IESFOgram.

References

1. L. Zhou, F. Duan, M. Corsar, F. Elasha, D. Mba, Applied Acoustics. **147**, 4-14 (2019)
2. R. Randall, N. Sawalhi, Rot. Machinery, Struct. Health Moni,, Shock and Vibr., **5**, 451- 461 (2011)

3. W. A. Smith, Z. Fan, Z. Peng, H. Li, R. Randall, MSSP, **75**, 371-394 (2016)
4. Z. Chen, A. Mauricio, W. Li, K. Gryllias, MSSP, **140**, 106683 (2020)
5. L. Wen, X. Li, L. Gao, Y. Zhang ,IEEE Trans. Ind. Electron., **65**, 5990-5998 (2018),
6. R. Randall, J. Antoni, J. Chobsaard, MSSP, **15**, 945-962 (2001)
7. D. Abboud, S. Baudin, J. Antoni, D. Remond, M. Eltabach, O, Sauvage, MSSP, **75**, 280- 300 (2016)
8. A. Mauricio, J. Qi, K. Gryllias, Proceedings of ASME TurboExpo (2018)
9. A. Mauricio, W. Smith, R. Randall, J. Antoni, K. Gryllias, Proc. ISMA, 17-19 Sep. (2018)
10. H. Cheng, Y. Zhang, W. Lu, Z. Yang, Measurement, **143**, 155-179 (2019)